The Agricultural Tenancies Act 1995

AUSTRALIA
The Law Book Company
Brisbane ● Sydney ● Melbourne ● Perth

CANADA
Carswell
Ottawa ● Toronto ● Calgary ● Montreal ● Vancouver

AGENTS:
Steimatzky's Agency Ltd., Tel Aviv;
N. M. Tripathi (Private) Ltd., Bombay;
Eastern Law House (Private) Ltd., Calcutta;
M.P.P. House, Bangalore;
Universal Book Traders, Delhi;
Aditya Books, Delhi;
MacMillan Shuppan KK, Tokyo;
Pakistan Law House, Karachi

The Agricultural Tenancies Act 1995

with annotations by

Della Evans

*Partner, Burges Salmon,
Solicitors, Bristol*

LONDON
SWEET & MAXWELL
1995

Published in 1995 by
Sweet & Maxwell Limited of
South Quay Plaza,
183 Marsh Wall, London
Typeset by MFK Information Services Ltd.,
Hitchin, Herts
Printed and bound in Great Britain by
Butler & Tanner Ltd., Frome and London

**A CIP catalogue record for this book is available
from The British Library**

ISBN 0–421–54590–9

All rights reserved.
U.K. statutory material in this publication is acknowledged as Crown copyright.
No part of this publication may be reproduced or transmitted in any form or by any means, or stored in any retrieval system of any nature, without prior written permission, except for permitted fair dealing under the Copyright, Designs and Patents Act 1988, or in accordance with the terms of a licence issued by the Copyright Licensing Agency in respect of photocopying and/or reprographic reproduction. Application for permission for other use of copyright material including permission to reproduce extracts in other published works shall be made to the publishers. Full acknowledgement of author, publisher and source must be given.

© *Sweet & Maxwell 1995*

CONTENTS

References are to page numbers

Table of Cases	ix
Table of Statutes	xi

Agricultural Tenancies Act 1995

Part I:	General Provisions	8–4
Part II:	Rent Review Under Farm Business Tenancy	8–19
Part III:	Compensation on Termination of Farm Business Tenancy	8–27
Part IV:	Miscellaneous and Supplemental	8–40
Schedule		8–52
Index		8–61

PREFACE

The Agricultural Tenancies Act 1995 revolutionises the agricultural landlord and tenant relationship and must be seen as the most significant change in agricultural holdings law since the Agricultural Holdings Act 1948.

The Act follows closely an industry agreement reached between the Country Landowners' Association, the Tenant Farmers' Association, the National Farmers' Union and the National Federation of Young Farmers Clubs. Clearly, any Act which results from an agreement between those organisations is a compromise and this is obvious from the tension in the Act between freedom of contract and regulation of the landlord and tenant relationship. However, there is no doubt that it is a significant achievement that industry agreement was reached at all.

The Agricultural Holdings Act 1986, which will continue to apply to existing lettings, heavily regulates and prescribes the nature of the landlord and tenant relationship and provides the agricultural tenant with, effectively, lifetime security of tenure. The new Act starts from the premise of freedom of contract whereby the parties can agree most of their own terms and the method by which they require disputes to be resolved. There is, however, regulation in the context of the length of notices to quit, rent review, removal of tenants' fixtures and compensation for improvements.

Land agents and other professionals used to drawing up agricultural tenancies under the Agricultural Holdings Act 1986, which are predominantly annual periodic tenancies and where the 1986 Act effectively provides many of the clauses in the tenancy agreement, will have a major task in adjusting to the new legislation. Very much more is left to negotiation between the parties, and the land agent's or solicitor's role in negotiating and in concluding that agreement is therefore much more significant. In particular, both solicitors and land agents will have to be far more aware of general landlord and tenant and general property law than they ever had to be when operating under the Agricultural Holdings Act 1986. It is hoped that this handbook will assist in the learning process which now has to be undertaken by everyone involved in the letting of agricultural land.

Much of the analysis in this handbook was assisted by the Burges Salmon Roadshow of lectures on the Act which we undertook in the second half of May this year lecturing at 10 venues around the country during the day and assisting the NFU in their Roadshows in the evening. The stimulating questions from the audience have certainly assisted in our understanding of the Act. However, I would like to thank in particular Peter Williams and Andrew Densham of Burges Salmon for their encouragement and assistance and also Jane Adderley at the NFU for answering some tricky questions. The staff at Sweet & Maxwell have, of course, been patient and understanding.

Della Evans
Burges Salmon
Narrow Quay House
Prince Street
Bristol BS1 4AH

June 29, 1995

TABLE OF CASES

References are to the General Note to the specified section and to the Introduction

Agricultural Mortgage Corp. v. Woodward [1995] 04 E.G. 155, C.A. s.31
Buckingham County Council v. Gordon (1986) 279 E.G. 853; [1986] 2 E.G.L.R. 8 .... s.10
City of London Corp. v. Fell [1994] 1 A.C. 458, H.L.; [1993] Q.B. 589; [1993] 2
    W.L.R. 710; [1993] 2 All E.R. 449; 91 L.G.R. 151; (1992) 65 P. &. C.R. 229;
    [1993] 04 E.G. 115, C.A. s.5
Dallhold Estates (U.K.) Pty (In administration) v. Lindsey Trading Properties
    [1994] 17 E.G. 148; *The Times*, December 15, 1993, C.A. s.36
Dow Agrochemicals v. Lane (E.A.) North Lynn (1965) 192 E.G. 737, Cty. Ct. s.1
Duke of Wellington's Parliamentary Estates, *Re*; King v. Wellesley [1972] Ch. 374 .... s.33
Elwes v. Maw [1802] 3 East 38 .... s.8
Faulks v. Faulks [1992] 1 E.G.L.R. 9; [1992] 15 E.G. 82 .... s.15
Fredco Estates v. Bryant [1961] 1 W.L.R. 76; 105 S.J. 86; [1961] 1 All E.R. 34 .... s.4
Gladstone v. Bower [1960] 2 Q.B. 384; [1960] 3 W.L.R. 575; 104 S.J. 763; [1960] 3 All
    E.R. 353; 58 L.G.R. 313, C.A.; affirming [1960] 1 Q.B. 170; [1959] 3 W.L.R.
    815; 103 S.J. 835; [1959] 3 All E.R. 475; 58 L.G.R. 75; [1959] C.L.Y. 27 ..... Intro, ss. 1, 5
Greenway v. Tempest (1983) unreported s.11
Hickson and Welch v. Cann (1977) 40 P. & C.R. 218, C.A. s.1
Jelley v. Buckman [1974] Q.B. 488 .... s.11
Johnson v. Moreton [1980] A.C. 37; [1978] 3 W.L.R. 583; (1978) 122 S.J. 697; [1978] 3
    All E.R. 37; (1978) 37 P. & C.R. 243; (1978) 247 E.G. 895, H.L. s.8
Land Securities v. Westminster City Council [1993] 1 W.L.R. 286; [1993] 4 All E.R.
    124; (1992) 65 P. & C.R. 387; [1992] 44 E.G. 153; [1992] N.P.C. 125 .... s.13
Mann v. Gardner (1990) 61 P. & C.R. 1; [1991] 11 E.G. 107; *The Times*, June 22,
    1990, C.A. s.10
Pooles case (1703) 1 SALK 368 .... s.8
Premier Dairies v. Garlick [1920] 2 Ch. 17 .... s.8
Puncknowle Farms v. Kane [1985] 3 All E.R. 790; [1986] 1 C.M.L.R. 27; [1985] 2
    E.G.L.R. 8; (1985) 275 E.G. 1283 .... s.17
Ramsey v. McClaren [1936] S.L.T. 35 .... s.28
Rhodes v. Dalby [1971] 1 W.L.R. 1325; 115 S.J. 623; [1971] 2 All E.R. 1144; (1971) 23
    P. & C.R. 309 .... s.31
Rutherford v. Maurer [1962] 1 Q.B. 16; [1961] 3 W.L.R. 5; 105 S.J. 404; [1961] 2 All
    E.R. 775, C.A. s.1
Saunders Trustees v. Ralph [1993] 28 E.G. 127 .... s.4
Styles v. Farrow (1977) 241 E.G. 623 .... s.11
Weinbergs Weatherproofs v. Radcliffe Paper Mill Co. [1958] Ch 437; [1958] 2
    W.L.R. 1; 102 S.J. 15; *sub nom.* Bleachers' Association's Leases, *Re*; Wein-
    bergs Weatherproofs v. Radcliffe Paper Mill Co. [1957] 3 All E.R. 663 .... s.5

TABLE OF STATUTES

References are to the General Note to the specified section or Part

1925	Settled Land Act 1925 (c. 18) s.33		1986	Agricultural Holdings Act 1986—*cont.*
	s.42 .... s.34			s.37(1)(a) s.4
	s.73 .... s.33			(b) s.4
	ss.83–89 .... s.33			(2) s.4
	s.84 .... s.33			s.39 .... s.4
	Sched. 3 .... s.33			s.45 .... s.4
	Law of Property Act 1925 (c. 20)—			s.46 .... s.4
				s.53 .... s.4
	s.28 .... s.34			s.65 .... s.15
	s.52 .... ss.1, 34, 35			s.74 .... s.24
	s.54 .... ss.34, 35			s.75 .... s.25
	s.99 .... ss.31, 34			s.79(2)–(5) s.4
	(3)(i) s.31			s.80(3)–(5) s.4
	(5)–(7) s.31			s.84 .... s.28
	(8) s.31			(1) s.28
	(13) s.31			s.88 .... s.32
	(13A) s.31			s.89 .... s.33
	s.140 .... ss.5, 11			s.96 .... Intro, s.1
	(2) ss.5, 24, 25			Pt. IV s.4
	s.149(6) s.7			Sched. 2 .... ss.9, 12, 13
1948	Agricultural Holdings Act 1948 (c.63) .... Intro			para. 4(2) s.10
1950	Arbitration Act 1950 (c.27)—			Sched. 3 .... Intro, s.6
				Sched. 7 .... s.33
	s.4(1) s.28			Sched. 8 .... s.15
	s.31 .... s.28			Sched. 11 .... Intro, s.28
1954	Landlord and Tenant Act 1954 (c.56) .... Intro			para. 7 .... s.28
				para. 14 .... s.28
	s.38 .... s.7			Sched. 14, para. 12 .... s.31
	Pt. II ss.1, 5, 7, 23, 31			Insolvency Act 1986 (c.45)—
1974	Solicitors Act 1974			
	s.22 35			s.423 .... s.31
	(2) s.35			Agriculture Act 1986 (c.49) s.16
1976	Agriculture (Miscellaneous Provisions) Act 1976 (c.55) .... Intro, s.4		1987	Landlord and Tenant Act 1987 (c.31)—
				ss.47, 48 .... s.36
1979	Arbitration Act 1979 (c.42)—		1988	Housing Act 1988 (c.50) s.1
			1995	Finance Act 1995 (c.4)
	s.2 .... s.28			s.155 .... Intro
	(2) s.28			Agricultural Tenancies Act 1995 (c.8) .... s.13, Pt. III
	s.5 .... s.28			s.1 .... ss.5, 23
1984	Agricultural Holdings Act 1984 (c.41) .... Intro, s.4			(4) s.36
1986	Agricultural Holdings Act 1986 (c.5) .... Intro, ss.1, 2, 4, 5, 6, 8, 10, 13, 15, 16, 22, 28, 31, 35			s.4 .... s.2
				s.5 .... ss.1, 6, 24
				(3) ss.6, 7
	s.1(3) s.1			s.6 .... s.5
	s.2 .... ss.1, 4			s.8 .... s.15
	s.10 .... s.8			(2)(d) s.17
	s.12 .... ss.9, 12, 13			(3), (4) s.17
	s.37 .... s.4			s.9 .... s.4, Pt. II, ss.10, 11, 12

xi

TABLE OF STATUTES

1995	Agricultural Tenancies Act 1995—*cont.*		1995	Agricultural Tenancies Act 1995—*cont.*	
	s.10	s.12		s.20(2)	ss.15, 21
	(4)	s.9		s.23	s.16
	s.12	ss.9, 29		s.24	s.25
	(b)	Pt. II, ss.9, 10, 29		(2)	s.25
	s.13	Pt. II, ss.10, 12		s.28	ss.12, 19, 22, 29
	(3)	s.20		(3)	s.30
	s.15	Intro, ss.8, 13, Pt. III		(5)	ss.12, 19, 22, 29
	(1)	s.20		s.29	ss.12, 19, 22, 28
	s.16	ss.8, 23		s.30	ss.12, 19, 28
	s.17	ss.8, 19		s.31	s.34
	(1)	s.19		s.36	ss.1, 5, 6, 10
	(2)	s.19		s.38	ss.1, 2, 4, 10
	s.18	s.20		(1)	ss.1, 23
	s.19	s.8		(2)	Intro, s.1
	(1)(a)	s.18		(4)	s.1
	(2)	ss.15, 17		(5)	s.8
	(8)	s.17		Pt. III	ss.8, 26
	(10)	ss.15, 17		Sched., para. 10	s.1
	s.20	ss.16, 18		para. 32	s.4

AGRICULTURAL TENANCIES ACT 1995*

(1995 c. 8)

ARRANGEMENT OF SECTIONS

PART I

GENERAL PROVISIONS

Farm business tenancies

SECT.
1. Meaning of "farm business tenancy".
2. Tenancies which cannot be farm business tenancies.
3. Compliance with notice conditions in cases of surrender and re-grant.

Exclusion of Agricultural Holdings Act 1986

4. Agricultural Holdings Act 1986 not to apply in relation to new tenancies except in special cases.

Termination of the tenancy

5. Tenancies for more than two years to continue from year to year unless terminated by notice.
6. Length of notice to quit.
7. Notice required for exercise of option to terminate tenancy or resume possession of part.

Tenant's right to remove fixtures and buildings

8. Tenant's right to remove fixtures and buildings.

PART II

RENT REVIEW UNDER FARM BUSINESS TENANCY

9. Application of Part II.
10. Notice requiring statutory rent review.
11. Review date where new tenancy of severed part of reversion.
12. Appointment of arbitrator.
13. Amount of rent.
14. Interpretation of Part II.

PART III

COMPENSATION ON TERMINATION OF FARM BUSINESS TENANCY

Tenant's entitlement to compensation

15. Meaning of "tenant's improvement".
16. Tenant's right to compensation for tenant's improvement.

Conditions of eligibility

17. Consent of landlord as condition of compensation for tenant's improvement.
18. Conditions in relation to compensation for planning permission.
19. Reference to arbitration of refusal or failure to give consent or of condition attached to consent.

Amount of compensation

20. Amount of compensation for tenant's improvement not consisting of planning permission.
21. Amount of compensation for planning permission.
22. Settlement of claims for compensation.

Supplementary provisions with respect to compensation

23. Successive tenancies.
24. Resumption of possession of part of holding.

---

* Annotations by Della Evans, Partner, Burges Salmon, Solicitors, Bristol

25. Compensation where reversionary estate in holding is severed.
26. Extent to which compensation recoverable under agreements.
27. Interpretation of Part III.

Part IV

Miscellaneous and Supplemental

Resolution of disputes

28. Resolution of disputes.
29. Cases where right to refer claim to arbitration under section 28 does not apply.
30. General provisions applying to arbitrations under Act.

Miscellaneous

31. Mortgages of agricultural land.
32. Power of limited owners to give consents etc.
33. Power to apply and raise capital money.
34. Estimation of best rent for purposes of Acts and other instruments.
35. Preparation of documents etc. by valuers and surveyors.

Supplemental

36. Service of notices.
37. Crown land.
38. Interpretation.
39. Index of defined expressions.
40. Consequential amendments.
41. Short title, commencement and extent.

Schedule:
—Consequential amendments.

An Act to make further provision with respect to tenancies which include agricultural land. [9th May 1995]

Parliamentary Debates

Hansard, H.L. Vol. 559, cols. 25, 485, 1089, 1112, 1166, 1205, Vol. 560, cols. 862, 884, 1258, 1279, Vol. 563, col. 1348. H.C. Vol. 254, col. 23, Vol. 258, col. 230.

Introduction and General Note

This Act reforms the law relating to the letting of agricultural land, providing an entirely new framework for the agricultural landlord and tenant relationship for lettings in England and Wales. The Act is not retrospective and will only apply to new lettings beginning (within the meaning of the Act) on or after September 1, 1995 when the Act comes into force.

Background

Since 1875 there has been legislation regulating the relationship between landlords and tenants of agricultural land culminating in legislation which provided the tenant with, effectively, lifetime security of tenure. The high watermark of statutory interference with freedom of contract came in 1976 with the retrospective introduction of succession rights upon the death of a tenant of an agricultural holding (the Agriculture (Miscellaneous Provisions) Act 1976 (c. 55)), although automatic rights of succession were abolished in the Agricultural Holdings Act 1984 (c. 41) for lettings after July 12, 1984.

The current agricultural holdings legislation, which will continue to apply to all existing lettings and to new lettings beginning before September 1, 1995, is the Agricultural Holdings Act 1986 (c. 5) which is an Act consolidating 10 previous Acts and following the scheme, so far as security of tenure is concerned, of the Agricultural Holdings Act 1948 (c. 63).

Security of tenure under the 1986 Act is afforded by controls over both the length of, and the operation of, notices to quit served by the landlord in relation to annual periodic tenancies and the conversion of other types of tenancy and contractual licences conferring exclusive possession into annual periodic tenancies. If a 1986 Act tenant objects to the notice to quit and complies with the statutory requirements, the consent of the Agricultural Land Tribunal is necessary before a landlord can rely on the notice to quit save in very limited circumstances set out in Sched. 3 to the 1986 Act. In very many cases, in the absence of default by the tenant, the controls upon the operation of notices to quit have led to lifetime security of tenure for the tenant with a consequent depreciation in the value of the landlord's land by up to 50 per cent.

Whilst the security of tenure of the 1986 Act and its statutory predecessors achieved a great deal for tenants, there has been increasing reluctance on the part of landowners to let under the 1986 Act regime as a result of the lack of flexibility and their inability to regain vacant possession of the land. The figures set out in the Ministry of Agriculture, Food and Fisheries (MAFF) Consultation Paper on the Reform of Agricultural Tenancy Law (1991) show that whilst in 1910, 90 per cent of agricultural land was tenanted, that figure had fallen to 36 per cent by 1991 (see also *Agricultural Land Tenure in England and Wales*, Winter, Richardson, Short and Watkins, research by the Centre for Rural Studies, Cirencester (published by the RICS in 1990)).

A variety of alternative devices have been utilised by land owners and farmers to achieve the farming of agricultural land without security of tenure either utilising exceptions within the 1986 Act (for example short-term grazing agreements of less than one year), exploiting loopholes within the 1986 Act (for example *Gladstone v. Bower* ([1960] 2 Q.B. 383) agreements of more than one but less than two years' duration which do not convert to annual periodic tenancies) or by avoiding a landlord and tenant relationship altogether by creating sharefarming, partnership or contracting arrangements (for further details see *Scammell & Densham's Law of Agricultural Holdings* (7th Edition, Butterworths) or Muir Watt, *Agricultural Holdings* (13th Edition, Sweet & Maxwell)). This was clearly an unsatisfactory state of affairs: short-term arrangements are not conducive to good husbandry and trying to force structures, such as partnerships, to work where essentially the parties wanted to be in a landlord and tenant relationship has resulted in difficulties.

The solution to the problem was perceived to be the deregulation of the landlord and tenant relationship leaving the parties free to negotiate their own letting arrangements. The road which led to the 1995 Act started with the 1991 consultation paper referred to above, where the objectives of the reform were stated to be to deregulate and simplify the existing legislation, to encourage the letting of land, and to provide an enduring framework of legislation which could accommodate change within the industry. In this initial consultation paper, it was envisaged that there would be almost no statutory interference with freedom of contract: there was to be no statutory provision for notices to quit or rent review or for dispute resolution, and compensation for improvements was only to be available as a fall back if the tenancy agreement was silent on the issue.

By the time of the MAFF detailed proposals paper in September 1992 (*Reform of Agricultural Holdings Legislation: Detailed Proposals*), the compromise between regulation and freedom of contract had already begun. It was recognised that the common law notice period for annual periodic tenancies (six months terminating on the term date) was too short in the context of the agricultural year and that there was some benefit in requiring fixed-term tenancies of a certain duration to be brought to an end by notice rather than simply by effluxion of time. There was to be compensation for tenants' improvements and a right for the parties to have the rent reviewed on the basis of an open market rent unless the parties chose their own basis in the tenancy agreement or stated that there was not to be a rent review. The emphasis had shifted away from leaving the parties to the court system for dispute resolution and towards encouraging alternative dispute resolution with a fall back of statutory arbitration. Industry agreement was sought to the proposals and a compromise was achieved which was set out in an industry agreement of 1993 backed by the Country Landowners' Association, Tenant Farmers' Association, the National Farmers' Union and the National Federation of Young Farmers' Clubs. The industry agreement is largely reflected in the provisions of the Act itself.

Main features of the Act

(1) The parties are free to negotiate the term of their tenancy. There is no minimum term. The hope is that long fixed-term tenancies will result from this legislation which will to an extent compensate the tenant for the lack of security of tenure but will allow the landlord to know exactly when he can get the land back.

(2) There is minimal security of tenure linked only to a length of notice requirement reflecting the need to allow the tenant time in the context of the farming year to vacate the land. If a landlord complies with the length of notice requirements set out in the Act, the tenant has no response that he can make to that notice provided that it is valid in accordance with the common law requirements and leaves the average tenant in no doubt as to what is required of him. A landlord does not have to wait until the tenant defaults before he can serve a notice to quit. However, there are disadvantages for a landlord letting on a fixed-term tenancy who would have let on an annual tenancy under the 1986 Act. Under the 1986 Act, for example, if the tenant fails to pay his rent, the landlord can serve a notice to pay under Case D of Sched. 3 to the 1986 Act requiring the tenant to pay that rent within two months. If the tenant fails to pay the rent within two months, the landlord can serve a notice to quit to which the tenant has no answer. Under the new Act, if a tenant under a fixed-term tenancy fails to pay his rent, the landlord is dependent upon the common law remedies. He can sue or distrain or he can make the tenant bankrupt.

However, if he wishes to regain possession of the farm as a result of the non-payment of rent, he must (if he can) forfeit the lease. As the tenant is then able to apply for relief from forfeiture, this does not guarantee, in the way that the 1986 Act guaranteed, that the landlord will regain possession of the farm (for a detailed analysis of forfeiture see Woodfall, *Landlord and Tenant*, Vol. 1, Chap. 17 (Sweet & Maxwell)).

(3) In recognition of the realities of life in the rural community, the Act provides for the possibility of substantial diversification into non-agricultural activities without the danger of the tenancy slipping out of the farm business tenancy regime and into the Landlord and Tenant Act 1954 (c. 56), Part II, as a general commercial letting.

(4) The landlord and tenant have a limited number of options available to them as far as rent review is concerned. If they fail to choose one of those options, the Act provides for the rent to be reviewed on an open market basis. The parties are free to choose the frequency of reviews but, if they fail to do so, the reviews will be every three years.

(5) The tenant is free in most cases to remove fixtures.

(6) The tenant will be compensated for tenants' improvements (as defined in s.15) where the landlord's written consent has been obtained to those improvements or, in certain cases, the approval of an arbitrator has been obtained.

(7) The emphasis, so far as dispute resolution is concerned, is on encouraging the parties to choose their own alternative dispute resolution mechanism but with the fall back of statutory arbitration should the parties either not choose their own mechanism or should one of the parties not wish to use that mechanism in connection with any particular dispute. Arbitrations are under the Arbitration Acts 1950–1979 and there is no discrete arbitration regime as there is in Sched. 11 to the 1986 Act.

(8) The Act makes no provisions for repairs or for dilapidation claims or for compensation other than in relation to tenants' improvements as defined by s.15. Much more is left for negotiation between the parties.

As the Act passed through first the House of Lords, and then the House of Commons, it was clear that there were two stumbling blocks in the way of its success, despite survey reports from the RICS that the farm business tenancy regime would lead to a significant increase in agricultural land available for letting (see *Farm Business Tenancies New Farms and Land 1995 to 1997*, RICS, October 1994). First, there was the lack of commitment from the Labour Party that, should they get into power, they would not introduce retrospective legislation conferring security of tenure upon tenants under farm business tenancies and, secondly, the inheritance tax treatment of let land was still less advantageous. Both of those stumbling blocks have been removed. The Labour Party has given a commitment that it will not introduce legislation which will retrospectively confer security of tenure on tenants holding under farm business tenancies (see *Hansard*, H.C. Vol. 258, cols. 256–257). In addition, an amendment has been made in the Finance Act 1995 (c. 4) (s.155) so that 100 per cent agricultural property relief will be allowed in respect of land let on or after September 1, 1995.

The Act was given Royal Assent on May 9, 1995 and comes into force on September 1, 1995.

ABBREVIATIONS
MAFF: Ministry of Agriculture, Fisheries and Food.
The 1986 Act: The Agricultural Holdings Act 1986.
WOAD: Welsh Office Agriculture Department.

PART I

GENERAL PROVISIONS

Farm business tenancies

Meaning of "farm business tenancy"

1.—(1) A tenancy is a "farm business tenancy" for the purposes of this Act if—
  (a) it meets the business conditions together with either the agriculture condition or the notice conditions, and
  (b) it is not a tenancy which, by virtue of section 2 of this Act, cannot be a farm business tenancy.

(2) The business conditions are—
  (a) that all or part of the land comprised in the tenancy is farmed for the purposes of a trade or business, and
  (b) that, since the beginning of the tenancy, all or part of the land so comprised has been so farmed.

(3) The agriculture condition is that, having regard to—
(a) the terms of the tenancy,
(b) the use of the land comprised in the tenancy,
(c) the nature of any commercial activities carried on on that land, and
(d) any other relevant circumstances,
the character of the tenancy is primarily or wholly agricultural.
(4) The notice conditions are—
(a) that, on or before the relevant day, the landlord and the tenant each gave the other a written notice—
(i) identifying (by name or otherwise) the land to be comprised in the tenancy or proposed tenancy, and
(ii) containing a statement to the effect that the person giving the notice intends that the tenancy or proposed tenancy is to be, and remain, a farm business tenancy, and
(b) that, at the beginning of the tenancy, having regard to the terms of the tenancy and any other relevant circumstances, the character of the tenancy was primarily or wholly agricultural.
(5) In subsection (4) above "the relevant day" means whichever is the earlier of the following—
(a) the day on which the parties enter into any instrument creating the tenancy, other than an agreement to enter into a tenancy on a future date, or
(b) the beginning of the tenancy.
(6) The written notice referred to in subsection (4) above must not be included in any instrument creating the tenancy.
(7) If in any proceedings—
(a) any question arises as to whether a tenancy was a farm business tenancy at any time, and
(b) it is proved that all or part of the land comprised in the tenancy was farmed for the purposes of a trade or business at that time,
it shall be presumed, unless the contrary is proved, that all or part of the land so comprised has been so farmed since the beginning of the tenancy.
(8) Any use of land in breach of the terms of the tenancy, any commercial activities carried on in breach of those terms, and any cessation of such activities in breach of those terms, shall be disregarded in determining whether at any time the tenancy meets the business conditions or the agriculture condition, unless the landlord or his predecessor in title has consented to the breach or the landlord has acquiesced in the breach.

DEFINITIONS
"agriculture": s.38(1).
"agricultural": s.38(1).
"agricultural condition": subs. (3).
"beginning of the tenancy": s.38(4).
"business conditions": subs. (2).
"farm business tenancy": ss.1, 2.
"farmed": s.38(2).
"gave": s.36.
"landlord": s.38(1), (5).
"notice conditions": subs. (4).
"relevant day": subs. (5).
"tenancy": s.38(1).
"tenant": s.38(1), (5).

GENERAL NOTE
The provisions of this Act apply only to farm business tenancies. Section 1 defines a "farm business tenancy" and the definition is by reference to compliance with the conditions set out in s.1. Proof of compliance is assisted by a presumption (see subs. (7)) and by a disregard for unlawful uses (see subs. (8)).

The first note of importance is that the Act applies only to *tenancies* which comply with the s.1 conditions: there is no equivalent to s.2 of the 1986 Act which converts informal licence arrangements into tenancies. There is no requirement that a farm business tenancy be in writing and the tenancy can, therefore, be oral, written or by deed, provided that the general formalities required by property law are complied with in relation to the particular term granted. Leases for a term exceeding three years and other leases not taking effect in possession and/or not at the best rent must be contained in a deed (see s.52 of the Law of Property Act 1925 (c. 20)). There is no minimum term for which a farm business tenancy must be granted. It can, therefore, be for a fixed term of anything from, for example, one day to 999 years or a periodic tenancy of any period. "Tenancy" is defined by s.38(1) (see below) to include a sub-tenancy or an agreement for a tenancy or a sub-tenancy.

Under the 1986 Act, as we have already seen (see the Introduction above) many short-term letting arrangements were designed to avoid the lifetime security of tenure of that Act. Either as a result of specific exceptions in the Act (for example, certain grazing lets) or because of loopholes (for example, *Gladstone v. Bower* ([1960] 2 Q.B. 384) tenancies for more than one but less than two years) a short-term arrangement could be made which would simply expire on its term date. The farm business tenancy is capable of replacing all of those short-term arrangements. Grazing agreements for a specified period of the year can now be by farm business tenancy as can a letting equivalent to the length of the letting under a *Gladstone v. Bower* arrangement. The landowner will be no worse off in respect of such short-term lettings than he was before this Act: short-term farm business tenancies for two years or less, will simply expire by effluxion of time (see s.5 below).

As this Act applies only to tenancies, it will still be possible to grant licences for grass keep without falling within the farm business tenancy regime, provided that the grant is properly a licence and does not confer exclusive possession on the grazier.

Subs. (1)

All farm business tenancies must meet the business conditions (see subs. (2) below). There is then a choice which is in the hands of the parties at the beginning of the tenancy. If the parties choose not to serve the notices required by the notice conditions (see subs. (4) below) before the "relevant day" (see subs. (5) below), the tenancy will have to comply with the agriculture condition to be a farm business tenancy. The combination of the business conditions and the notice conditions allow the farmer to diversify substantially after the tenancy has been set up, to a point where commercial farming is a very minor part of the activities being carried on on the land. However, the legislation is designed to allow diversification of *farming* businesses and not to enable landlords to force a predominantly non-farming enterprise into this regime to avoid the greater security of tenure of the Landlord and Tenant Act 1954 (c. 56), Pt. II, which applies to general commercial lettings. This is prevented by a minimum requirement that all farm business tenancies must be primarily or wholly agricultural at the beginning of the tenancy. As any commercial enterprise which fails to comply with the s.1 conditions is likely to fall within the Landlord and Tenant Act 1954, Pt. II, landlords ought to be careful to ensure continuing compliance. This is not as difficult as it appears. For fixed-term leases, it may be possible as a "belt and braces" exercise to exclude, by an agreement approved by the court, the security of tenure provisions of the 1954 Act (see s.38 of that Act). It is not yet clear to what extent the courts would be prepared to do this, because if the tenancy is a farm business tenancy at the outset, it is specifically excluded from the provisions of Pt. II of the 1954 Act (see para. 10 of the Schedule to this Act). Control is also possible through the user covenants in the tenancy itself (see subs. (8) below).

Subs. (2)

All farm business tenancies must comply with the business conditions. The business conditions require commercial farming on part of the land let on the farm business tenancy at all times from the beginning of the tenancy. The beginning of the tenancy is defined by reference to the date upon which the tenant is entitled to go into possession under the terms of the tenancy (see s.38(4)) and is not, therefore, necessarily the date the tenancy agreement was entered into. The reason is obvious: whilst the business conditions do not require the tenant personally to farm (although there may be constraints in the tenancy agreement itself), the tenant cannot be in control of the activities on a holding until he is entitled to go into possession.

There is a positive requirement in subs. (2)(b) that there be no break in compliance at any time during the life of the tenancy (*cf.* the agriculture condition in subs. (3) below). If there is a period when there is no commercial farming on the holding, however short and whatever the reasons and however far in the past, the status of the tenancy as a farm business tenancy will be at an end (although see the presumption in subs. (7) in relation to historic non-compliance). If there is still non-farming commercial activity on the holding, the letting is likely to fall within the security of tenure of the Landlord and Tenant Act 1954, Pt. II.

This section requires only *part* of the land to be farmed with no stated or specified percentage requirement. It is possible that the courts will impose a *de minimis* test but there is no requirement that the part farmed be a self-sufficient business or, indeed, that the business which the farming supports needs to be agricultural in nature (see below for the meaning of "farmed" and "agriculture"). There is no requirement that the part of the land being farmed remain the same throughout the term of the tenancy and it is possible to move the farming activity around the holding. Care will need to be taken to ensure that there is no period of non-compliance when no part of the land is being farmed whilst the farming activity is moved to another area of land.

The requirement that all or part of the land be *farmed*, as opposed to used for agriculture is new. There is no definition of "farmed", save that s.38(2) provides that it includes the carrying on in relation to land of any agricultural activity and "agriculture" is defined in s.38(1) by importing the definition found in s.96 of the 1986 Act (save for a change in relation to the definition of livestock). The choice of such a wide term is deliberate: it is sufficiently flexible to grow with changes and developments within the industry and does not (as yet) suffer from many years of judicial scrutiny as "agriculture" does.

There may be circumstances where an activity could be regarded as farming although it would not be agricultural. For example, growing crops for the purpose of testing pesticides has been held not to be "agriculture" (see *Dow Agrochemicals v. Lane (E.A.) (North Lynn)* (1965) 192 E.G. 737) although it could possibly be farming as could the rearing of livestock for research purposes. No assistance is given in the Act as to whether set-aside land can be said to be farmed.

As the definition of agriculture remains the same, grazing remains as an independent agricultural activity. As with the 1986 Act (see *Rutherford v. Maurer* [1962] 1 Q.B. 16) there does not appear to be any requirement that the trade or business which the farming activity supports be itself farming or agricultural in nature. Where the predominant use of land within a farm business tenancy is grazing, it does not matter that the business supported by that grazing is, for example, a riding school or a stud farm (see the cases on horses and grazing cited in *Scammell & Densham's Law of Agricultural Holdings*, pp. 27–29 (7th edition, Butterworths) and also see *The Agricultural Holdings Act 1986* by James Muir Watt, Current Law Statutes Annotated Reprints, at pp. 5–124 and 5–125 (Sweet & Maxwell)).

The word "farmed" is used *only* in the business conditions and not in the notice conditions or the agriculture condition where the word used is agriculture. Those farming activities which are not agricultural are therefore regarded as diversification activities under the Act and the tenancy must be primarily agricultural at the outset. If there is "farming" at the outset but no agricultural activity the tenancy cannot be a farm business tenancy.

"*For the purposes of a trade or business*". The requirement that the farming be commercial in nature is not new in concept. Hobby farming or the use of the land for recreation or amenity was also excluded from the 1986 Act. Essentially, the idea is to cut out non-commercial activities. In *Hickson and Welch v. Cann* (1977) 40 P. & C.R. 218, Bridge LJ said "There is all the difference in the world between what hundreds and hundreds of people all over the country do, that is, add a small amount to their earnings by buying or selling animals of one sort or another, and the carrying on of a trade or business of an agricultural nature". Growing vegetables primarily for own use, even if the surplus is sold or keeping a few sheep as pets, even if they are eventually sold, will not amount to using the land for a trade or business. Although grazing is farming, the use of land as a pony paddock or to keep horses used for hunting will not comply with the business conditions because the farming is not for the purposes or a trade or business.

Subs. (3)

If the parties do not choose to rely on the notice conditions in subs. (4) the agriculture condition must be complied with together with the business conditions.

There are two things to note about the agriculture condition. First, substantial diversification is not catered for as the requirement is that the character of the tenancy must be primarily or wholly agricultural. Secondly, non-compliance with the agriculture condition does not have the same consequences as non-compliance with the business conditions. Non-compliance with the agriculture condition will take the tenancy out of the farm business tenancy regime, but later compliance may bring it back in again. It is only really, therefore, at the time of testing the status of the tenancy as a farm business tenancy that the agriculture condition becomes relevant. If, at the time of testing, the tenancy is not primarily or wholly agricultural but is commercial in nature, the tenant may have the greater security of tenure of the business tenancy regime in the Landlord and Tenant Act 1954, Pt. II. In most cases therefore it is likely that professional advice would be to opt for the greater certainty and flexibility of the notice conditions. The agriculture condition is likely to be relied upon primarily where professional advice has not been sought and where, therefore, notices have not been exchanged or where defective notices or defective service of notices means that reliance cannot be placed on the notice conditions. If, however, the parties intend a relatively short-term agreement and are prepared to agree to very restrictive user covenants limiting use to an agricultural use, they may choose not to serve notices. Subsection (8) (see below) means that unlawful uses outside of the scope of the user covenants can be ignored in assessing compliance with the agriculture condition save where the landlord consents or acquiesces in the breach. In circumstances, for example, where grass keep is being auctioned and, therefore, the service of notices is not feasible, tight user covenants to grazing and/or mowing only, will suffice to ensure compliance with the agriculture condition.

"Primarily or wholly agricultural" is a new phrase. Under the 1986 Act, to establish that a contract of tenancy was for an agricultural tenancy it was necessary to show that the whole of the land subject to such exceptions as did not substantially affect the character of the tenancy was let for use as agricultural land. Whilst the wording in the new Act is different, the concept is familiar. Factors such as the percentage area of the holding used for agriculture; the percentage that agricultural and non-agricultural activities contribute to turnover and profit; the amount of time and labour which each activity takes up, will all be relevant and, at the end of the day, it will be a matter of overall impression. A non-agricultural activity which is insubstantial in terms of acreage but is predominant in terms of income may well mean in some cases that the tenancy is not primarily agricultural.

"Agriculture" is defined in s.38(1) and is the same definition as found in s.96 of the 1986 Act save for the fact that livestock is now differently defined. It should be noted that unlike the business conditions, the agriculture condition does not use the word "farmed". It is possible, therefore, for the entire holding to be farmed but not to be a farm business tenancy if the use is not agricultural.

Subs. (4)

An exchange of notices between landlord and tenant in the form set out in this subsection will allow substantial diversification away from agricultural activities without the danger of the tenancy losing its status as a farm business tenancy. Whilst the character of the tenancy must be primarily or wholly agricultural at the beginning of the tenancy (*i.e.* when the tenant is entitled to go into possession—see s.38), diversification thereafter into non-agricultural use will not breach the notice conditions. The notice conditions require matters to be looked at only at the beginning of the tenancy. Provided, therefore, that part of the land is always farmed for the purposes of a trade or business in compliance with the business conditions, the tenancy will remain a farm business tenancy despite possibly predominantly non-agricultural activity on the land.

The notice conditions are likely to be relied upon where diversification is intended by the tenant or to ensure certainty where diversification could happen unlawfully and where a landlord is concerned about consent or acquiescence in breach of the user covenants in longer term tenancies (see subs. (8) below), or where diversification, whilst not intended by the tenant, is permissible under the terms of user covenants which have been drafted widely in order not to impact adversely on the rent. The notice conditions still require the letting to be primarily or wholly agricultural at the outset to ensure that businesses which are intended to be predominantly non-agricultural cannot be farm business tenancies, thereby avoiding the security of tenure provisions of the Landlord and Tenant Act 1954, Pt. II. The purpose of this regime is to allow diversification of essentially farming businesses.

The phrase "primarily or wholly agricultural" has been discussed in connection with the agriculture condition (see subs. (3) above), the only difference is that here, as a snapshot picture is being taken at the beginning of the tenancy, greater emphasis will be placed, of necessity, on the terms of the tenancy itself.

There is no prescribed form for the notice to be served under this section although, as can be seen, there is prescribed information which must be included in the notice. For evidential purposes each notice should be signed, dated and acknowledged by the recipient. Care must be taken with service of the notice and reference should be made to s.36 and the General Note to that section. Section 36 sets out prescribed methods of service for any notice or document required or authorised to be given under the Act.

Failure to serve notices, the service of defective notices or defective service of notices does not necessarily mean that the tenancy will not be a farm business tenancy. If the agriculture condition is complied with, that will act as a safety net for most lettings.

Subs. (5)

This subsection defines the date by which the notices referred to above must be exchanged. Notices must be exchanged by the earlier of the date of a written tenancy agreement and the date upon which the tenant is entitled to go into possession. There is no requirement that notices be exchanged first on that day, before the tenancy is entered into.

Subs. (6)

The requirement for a separate notice not within the body of the tenancy agreement mirrors that for assured shorthold tenancies under the Housing Act 1988 (c. 50). It is considered more likely that a tenant will understand the importance of such a notice, read it thoroughly and obtain proper advice upon it if it is not just one of many clauses in a tenancy agreement.

Subs. (7)

Proof of compliance with the business conditions at the time when the tenancy's status as a farm business tenancy is challenged raises a presumption of historical compliance since the date upon which the tenant was entitled to go into possession. The burden of proof is therefore on the party trying to show historical non-compliance to prove that there has been a period, however short, when there was no commercial farming on the holding.

Subs. (8)

It is not possible for the tenant unilaterally to bring himself into or out of the farm business tenancy regime in breach of the terms of his tenancy agreement. Unlawful uses, prohibited by the terms of the tenancy, or failure to comply with positive obligations in the tenancy agreement in connection with user, can therefore be ignored when investigating ongoing compliance with the business conditions or the agriculture condition. A similar provision exists in s.1(3) of the 1986 Act.

Appropriately drawn user covenants, limiting the use of the holding to a specific agricultural use, to a use which is for agriculture in general or to a use which simply requires part of the land to be farmed commercially, will assist in ensuring compliance. However, restrictive user clauses may have an adverse impact on the rent achievable for the holding. In addition, in longer fixed-term tenancies, the risk of the landlord consenting to or acquiescing in the unlawful use means that controls through the user covenants alone are not enough to guarantee compliance.

A landlord will be taken to have acquiesced in a use in breach of covenant if from his conduct it can be inferred that he has waived his rights in relation to the breach. For example, if the tenant opens a garden centre from which the landlord regularly purchases plants. In many cases, the amount of time which has elapsed since the initial breach will be important, but there may be acquiescence without any delay at all.

Tenancies which cannot be farm business tenancies

2.—(1) A tenancy cannot be a farm business tenancy for the purposes of this Act if—
  (a) the tenancy begins before 1st September 1995, or
  (b) it is a tenancy of an agricultural holding beginning on or after that date with respect to which, by virtue of section 4 of this Act, the Agricultural Holdings Act 1986 applies.

(2) In this section "agricultural holding" has the same meaning as in the Agricultural Holdings Act 1986.

DEFINITIONS

"agricultural holding": subs. (2) and the 1986 Act, s.1.
"begins": s.38(4).
"farm business tenancy": ss.1, 2.
"tenancy": s.38(1).

GENERAL NOTE

The cut-off date between the farm business tenancy regime and the 1986 Act is the date of the coming into force of this Act on September 1, 1995. The legislation is not retrospective in any respect. It is by reference to the beginning of the tenancy, *i.e.* the date upon which the tenant is entitled to go into possession (see s.38) that the tenancy will fall either to be considered under the 1986 Act or to be considered under this Act. However, there are exceptions set out in s.4 (see below) which will allow certain tenancies created on or after September 1, 1995 to be protected under the 1986 Act.

Compliance with notice conditions in cases of surrender and re-grant

3.—(1) This section applies where—
  (a) a tenancy ("the new tenancy") is granted to a person who, immediately before the grant, was the tenant under a farm business tenancy ("the old tenancy") which met the notice conditions specified in section 1(4) of this Act,
  (b) the condition in subsection (2) below or the condition in subsection (3) below is met, and
  (c) except as respects the matters mentioned in subsections (2) and (3) below and matters consequential on them, the terms of the new tenancy are substantially the same as the terms of the old tenancy.

(2) The first condition referred to in subsection (1)(b) above is that the land comprised in the new tenancy is the same as the land comprised in the old tenancy, apart from any changes in area which are small in relation to the size of the holding and do not affect the character of the holding.

(3) The second condition referred to in subsection (1)(b) above is that the old tenancy and the new tenancy are both fixed term tenancies, but the term date under the new tenancy is earlier than the term date under the old tenancy.

(4) Where this section applies, the new tenancy shall be taken for the purposes of this Act to meet the notice conditions specified in section 1(4) of this Act.

(5) In subsection (3) above, "the term date", in relation to a fixed term tenancy, means the date fixed for the expiry of the term.

DEFINITIONS
  "farm business tenancy": ss.1, 2.
  "fixed term tenancy": s.38(1).
  "granted": s.38(3).
  "holding": s.38(1).
  "new tenancy, the": subs. (1)(a).
  "notice conditions": s.1(4).
  "old tenancy, the": subs. (1)(a).
  "tenancy": s.38(1).
  "tenant": s.38(1), (5).
  "term date, the": s.5(2).

GENERAL NOTE
  Where a farm business tenancy is granted with reliance being placed on the notice conditions, compliance with the notice conditions will have to be considered afresh in all cases of surrender and regrant. This is so whether what is intended by the parties is in fact a surrender and regrant or whether the parties are purporting to vary the terms of an existing farm business tenancy but where, in law, this amounts to a surrender and regrant. Without s.3, in all cases of surrender and regrant, whether express or inadvertent, new notices would have to be served and the parties would have to be able to show, in accordance with the second notice condition, that at the time of the regrant the character of the tenancy was primarily or wholly agricultural. With farms which have become highly diversified, the second condition could not be fulfilled and, where the surrender and regrant is inadvertent, the parties are unlikely to have considered the service of fresh notices.
  Section 3 prevents these problems from arising in two common situations: small changes in the area of land let and moving forward the term date in a fixed-term tenancy. In both cases, provided that the old tenancy met the notice conditions, and the new tenancy, apart from changes in the area of land or the term date, is on substantially the same terms, the fulfilment of the notice conditions at the outset of the old tenancy will stand for the new. No new notices need be served and no investigation is necessary into the primary use of the holding at the time of the new

grant. For all other situations of surrender and regrant, either the notice conditions must be complied with again or the parties will have to rely on the fallback of the agriculture condition. Care will need to be taken in agreeing variations which may amount to a surrender and regrant, particularly in relation to highly diversified estates.

Exclusion of Agricultural Holdings Act 1986

Agricultural Holdings Act 1986 not to apply in relation to new tenancies except in special cases

4.—(1) The Agricultural Holdings Act 1986 (in this section referred to as "the 1986 Act") shall not apply in relation to any tenancy beginning on or after 1st September 1995 (including any agreement to which section 2 of that Act would otherwise apply beginning on or after that date), except any tenancy of an agricultural holding which—
  (a) is granted by a written contract of tenancy entered into before 1st September 1995 and indicating (in whatever terms) that the 1986 Act is to apply in relation to the tenancy,
  (b) is obtained by virtue of a direction of an Agricultural Land Tribunal under section 39 or 53 of the 1986 Act,
  (c) is granted (following a direction under section 39 of that Act) in circumstances falling within section 45(6) of that Act,
  (d) is granted on an agreed succession by a written contract of tenancy indicating (in whatever terms) that Part IV of the 1986 Act is to apply in relation to the tenancy,
  (e) is created by the acceptance of a tenant, in accordance with the provisions as to compensation known as the "Evesham custom" and set out in subsections (3) to (5) of section 80 of the 1986 Act, on the terms and conditions of the previous tenancy, or
  (f) is granted to a person who, immediately before the grant of the tenancy, was the tenant of the holding, or of any agricultural holding which comprised the whole or a substantial part of the land comprised in the holding, under a tenancy in relation to which the 1986 Act applied ("the previous tenancy") and is so granted merely because a purported variation of the previous tenancy (not being an agreement expressed to take efect as a new tenancy between the parties) has effect as an implied surrender followed by the grant of the tenancy.

(2) For the purposes of subsection (1)(d) above, a tenancy ("the current tenancy") is granted on an agreed succession if, and only if,—
  (a) the previous tenancy of the holding or a related holding was a tenancy in relation to which Part IV of the 1986 Act applied, and
  (b) the current tenancy is granted otherwise than as mentioned in paragraph (b) or (c) of subsection (1) above but in such circumstances that if—
    (i) Part IV of the 1986 Act applied in relation to the current tenancy, and
    (ii) a sole (or sole surviving) tenant under the current tenancy were to die and be survived by a close relative of his,
  the occasion on which the current tenancy is granted would for the purposes of subsection (1) of section 37 of the 1986 Act be taken to be an occasion falling within paragraph (a) or (b) of that subsection.

(3) In this section—
  (a) "agricultural holding" and "contract of tenancy" have the same meaning as in the 1986 Act, and
  (b) "close relative" and "related holding" have the meaning given by section 35(2) of that Act.

DEFINITIONS
  "agreed succession": subs. (2).
  "agricultural holding": subs. (3) and the 1986 Act, s.1.

"beginning": s.38(4).
"close relative": subs. (3) and the 1986 Act, s.35(2).
"contract of tenancy": subs. (3) and the 1986 Act, s.1(5).
"current tenancy, the": subs. (2).
"Evesham custom": 1986 Act, s.80(3)–(5).
"holding": s.38(1).
"previous tenancy, the": subs. (1)(f).
"related holding": subs. (3) and the 1986 Act, s.35(2).
"tenancy": s.38(1).
"tenant": s.38(1), (5).

GENERAL NOTE
This section makes it clear that, subject to the limited exceptions set out below, the 1986 Act will not apply to tenancies which begin on or after September 1, 1995 or to informal licence arrangements which begin on or after that date which would otherwise be converted into tenancies by s.2 of the 1986 Act. A tenancy begins on the date upon which the tenant is entitled to go into possession (see s.38). This is not necessarily the date on the tenancy agreement.

With one exception (see subs. (1)(a)), the s.4 exceptions cover circumstances where, for various reasons, it can reasonably be said that the legitimate expectations of the tenant would be that he or his successors would have 1986 Act security and therefore the exceptions are within the spirit of the government's stated intention that this Act will not be retrospective.

There is no general ability (apart from subs. (1)(a)) to contract back into 1986 Act security, although this did appear, almost certainly inadvertently, in an earlier draft of the Bill. The exceptions in s.4 are limited and the idea is that 1986 Act tenancies should disappear as soon as possible, with as few new lettings on or after September 1, 1995 under that regime as accords with the new Act not being retrospective. It is recognised that this could cause difficulties, for example, where a landlord and a tenant wish the tenant to move farms on or after September 1, 1995. Whilst, to an extent, the lack of security of tenure can be compensated for by the grant of a long fixed-term, other constraints, such as a limited range of options in respect of rent review mechanisms (see s.9 *et seq*), make it extremely difficult to replicate a 1986 Act tenancy.

Subs. (1)(a)
This exception does allow a limited ability to contract back into 1986 Act security. It recognises that, in the run up to the coming into force of the legislation on September 1, 1995, parties may wish to take an opportunity to reorganise estates – moving tenants between farms – or to promise 1986 Act security to new tenants or effect express surrenders and regrants not covered by subs. (1)(e) below. However, the exception also recognises that it may not be convenient in terms of the farming year to move such tenants before September 1, 1995 which would create difficulties without this exception, as the cut-off date between the two Acts is by reference to the date upon which the tenant is entitled to go into possession.

This exception provides that if a written contract of tenancy (as defined by the 1986 Act) is entered into before September 1, 1995 stating that the 1986 Act is to apply to that tenancy, then it will apply regardless of the fact that the tenancy is not to begin within the meaning of the Act (*i.e.* the tenant is not to be entitled to go into possession) until on or after September 1. It will enable reorganisation plans to be put in place before September 1 with the tenants moved, say, at Michaelmas.

Subs. (1)(b)–(d)
This group of exceptions is designed to ensure that what might loosely be called succession tenancies, succeeding to tenancies currently governed by the 1986 Act, will themselves fall under the 1986 Act regime, to accord with the legitimate expectations of the parties. The Agriculture (Miscellaneous Provisions) Act 1976 (c. 55) first introduced a statutory entitlement to succession on the death of a tenant by a defined class of close relatives who satisfied certain eligibility tests. The Agricultural Holdings Act 1984 (c. 41) extended succession to retirement and altered the eligibility tests. The Agricultural Holdings Act 1986 (c. 5), which is the current legislation, now contains the succession rules. The basic position is that all 1986 Act tenancies carry the right to succession, save for those granted after July 12, 1984, which will only carry the right to succession if the tenancy itself is a succession tenancy or if the parties have contracted back into succession rights or if there is a new tenancy of substantially the same holding granted to an existing tenant whose original tenancy carried succession rights.

Subsection (1)(b) relates to succession tenancies obtained as a result of the direction of the Agricultural Lands Tribunal to an eligible person (s.39 of the 1986 Act) or to a nominated successor (s.53 of the 1986 Act).

Subsection (1)(c) relates to circumstances where the landlord has actually granted a tenancy to a successor following a direction but before the time when the direction would, in accordance with the provisions of s.45 and s.46 of the 1986 Act, entitle the successor to the new tenancy.

Subsection (1)(d) is the most problematic of this group of exceptions. Subsection (1)(d) recognises that there will be circumstances where a successor tenant is agreed between the parties and the matter never goes before the Agricultural Lands Tribunal. However, it was recognised by the drafters of the Act that simply to refer to an agreed succession without further definition could create an opportunity for contracting back into 1986 Act security simply by calling the new tenancy an agreed succession. There are two controls: the first is in subs. (1)(d) itself: the tenancy must state that Pt. IV of the 1986 Act (succession provisions) is to apply to the new tenancy. The second control is in the definition of an agreed succession set out in subs. (2) below.

Subs. (1)(e)

This exception is to encourage the ongoing working of the Evesham Custom. The custom, set out in the 1986 Act, allows investment by tenants in market gardens predominantly, though not exclusively, in and around Evesham in such a way as to enable a landlord not to have to pay compensation for the improvements in certain circumstances where the tenant quits the holding. Again, it is an exception to reflect the legitimate expectations of the parties and to ensure that this Act does not put tenants of market gardens in a position of not being able to obtain recompense for their investment on quitting the holding. The Evesham Custom is set out in s.80(3)–(5) of the 1986 Act. In brief, where the Agricultural Lands Tribunal is satisfied that a holding is suitable for the purposes of market gardening, it may direct that s.79(2)–(5) of the 1986 Act shall apply. Section 79(2)–(5) gives additional rights to tenants of market gardens in respect of improvements consisting of the planting of trees, bushes and certain plants or the erection, alteration or enlargement of buildings for the purposes of the trade or business of a market garden. If such a direction is made and a notice to quit is served by the tenant or the tenant becomes insolvent, the tenant will not obtain compensation from the landlord for the improvements unless he produces an offer in writing from a suitable person: (a) to accept a tenancy on the same terms and conditions; and (b) to pay all compensation due under the Act or the tenancy to the outgoing tenant; and the landlord has failed to accept that offer within three months.

Unless a new tenant were to obtain security of tenure under the 1986 Act, it is unlikely that one could be found who would be prepared to pay the outgoing tenant for the improvements. In those circumstances, the outgoing tenant would be left with a condition for obtaining compensation which he could not fulfil.

Subs. (1)(f)

This exception recognises that there are circumstances where parties will believe that they are merely varying the terms of a 1986 Act tenancy but where, as a matter of law, they have effected a surrender and regrant. If such a purported variation of a tenancy protected under the 1986 Act takes effect as a surrender, the grant of a new tenancy beginning on or after September 1, 1995 would, without this exception, put the tenant in a position of having lost his security of tenure without realising that that is what has happened. This exception goes some way to alleviating the problem. It is, however, limited in its scope and will not give 1986 Act protection to a new grant where: (1) there is an additional tenant on the new grant; (2) the land in the new grant is not the whole or a substantial part of the land in the old tenancy; (3) the arrangement is an express surrender and regrant and not an inadvertent surrender and regrant arising from a purported variation.

For circumstances which fall outside of this exception, care will need to be taken when advising tenants whether or not to accept variations to a 1986 Act tenancy after the coming into force of this Act. Ultimately, whether a purported variation operates as a surrender and regrant depends upon the intention of the parties. The addition of a new tenant is not necessarily a surrender and regrant (see *Saunders Trustees v. Ralph* [1993] 28 E.G. 127) and whilst the addition of extra land will often amount to a surrender and regrant, this is not invariably the case. In *Fredco Estates v. Bryant* [1961] 1 All E.R. 34, in the context of residential property, it was implied that if the regrant offered less security than the original grant, it could be said that the tenant would not have intended to put himself in that position and would not therefore have intended a surrender.

Subs. (2)

Reference should be made also to the commentary to subs. (1)(d) above. The intention of subs. (2) is to define "an agreed succession" in such a way as to ensure that it is not open to parties to use the exception in subs. (1)(d) as an indirect method of contracting back into 1986 Act security of tenure simply by calling the tenancy an agreed succession. The method by which subs. (2) attempts to achieve this is not straightforward. First, the tenancy to which the exception is to apply must be succeeding to a tenancy which does actually carry succession rights. The second condition is extremely convoluted and based on the application of hypotheses to the tenancy. What it amounts to is that a tenancy will only be treated as granted on an agreed succession for the purposes of this exception if the grant of the tenancy is in such circumstances that it would count as one of the two successions allowed by law under the 1986 Act.

Part of the reason for the complexity is the terms of the succession provisions in the 1986 Act. Instead of saying, in terms, that there shall be two successions, s.37 of the 1986 Act provides that there shall be no entitlement to apply to succeed if on two previous occasions when a sole or sole surviving tenant died, there had been a succession by direction or grant following direction or there had been an agreed succession by a close relative. This exception attempts to mirror that provision.

Whilst directions of the Agricultural Lands Tribunal and grants following directions are specifically excluded from falling within the definition of agreed succession having already been dealt with by other exceptions within s.4 (see subs. (1)(b) and (c) above), care has to be taken. The two circumstances which will count as a succession under the 1986 Act are set out in s.37(1)(a) and s.37(1)(b) of that Act. Reference is made to those provisions in s.4(2). While s.37(1)(a) refers to directions of the Agricultural Lands Tribunal and grants following directions, as a result of s.37(2) there are certain factual circumstances which will be deemed to fall within s.37(1)(a) which will not fall within subs. (1)(b) or (c) above. For example, s.37(2) provides that agreements prior to the date of the death of the tenant where a new tenant would be a close relative, if the old tenant died immediately before the grant, will be deemed to fall within s.37(1)(a) and be treated as a direction of the Agricultural Lands Tribunal. Section 37(1)(b) covers the situation where a new tenancy was granted to a close relative of a tenant who had died and who had become the sole remaining applicant for succession.

In determining eligibility for succession, the 1986 Act will still apply. However, in the Schedule to this Act, (para. 32), the 1986 Act is amended so that a potential successor does not have counted against him any land which he holds on a farm business tenancy of less than five years when assessing whether he is the occupier of a commercial unit of agricultural land which would disqualify him from being a person eligible for succession.

Termination of the tenancy

Tenancies for more than two years to continue from year to year unless terminated by notice

5.—(1) A farm business tenancy for a term of more than two years shall, instead of terminating on the term date, continue (as from that date) as a tenancy from year to year, but otherwise on the terms of the original tenancy so far as applicable, unless at least twelve months but less than twenty-four months before the term date a written notice has been given by either party to the other of his intention to terminate the tenancy.

(2) In subsection (1) above "the term date", in relation to a fixed term tenancy, means the date fixed for the expiry of the term.

(3) For the purposes of section 140 of the Law of Property Act 1925 (apportionment of conditions on severance of reversion), a notice under subsection (1) above shall be taken to be a notice to quit.

(4) This section has effect notwithstanding any agreement to the contrary.

DEFINITIONS
 "farm business tenancy": s.1.
 "fixed term tenancy": s.38(1).
 "given": s.36.
 "tenancy": s.38(1).

"term date, the": subs. (2).
"termination": s.38(1).

GENERAL NOTE
This section provides limited security of tenure to a tenant under a fixed-term farm business tenancy of more than two years by requiring positive action on the part of either the landlord or tenant to bring it to an end. Such a tenancy will not expire by effluxion of time. It can only be brought to an end on its term date by the service of a written notice in advance of the term date giving at least 12 but less than 24 months' notice. If such a notice is not served, the tenancy will continue as an annual periodic tenancy until brought to an end in accordance with the provisions of s.6 below.

It is possible that the courts will find such a continuation tenancy to be contractual in nature. Cases under the Landlord and Tenant Act 1954, Pt. II, where commercial business tenancies continue after the expiry of the fixed term, have concluded that the continuation tenancy is an extension of the contractual term with a variation as to the method of termination. See, for example, *Weinbergs Weatherproofs v. Radcliffe Paper Mill Co.* [1958] Ch 437 and *City of London Corp. v. Fell* [1993] QB 589 (C.A.) and [1994] 1 AC 458 (H.L.). This means that the tenant has an estate in land rather than simply a personal right to occupy. There is, however, a distinction between the two Acts: the 1954 Act does not refer to the tenancy continuing as an annual period tenancy. Section 5 of this Act, on the other hand, specifically provides that the farm business tenancy continues as a tenancy from year to year.

No contracting out is permitted from this section. Any provisions in the tenancy agreement for no notice or for a short notice will therefore be ineffective, although an agreed surrender between landlord and tenant remains a possibility.

Fixed-term tenancies with a duration of two years or less are not governed by the Act and will simply come to an end by effluxion of time on their term date. These tenancies (see the General Note to s.1 above) will replace the short-term grazing agreements, *Gladstone v. Bower* agreements and Ministry consent tenancies currently used to avoid the security of tenure of the 1986 Act.

Service of all notices and documents authorised or required to be served under this Act must be served in accordance with the provisions of s.36 (see below). There is, however, no prescribed form for the notice, although it must comply with the common law requirements.

Subs. (3)
Section 140 of the Law of Property Act 1925 deals with the apportionment of conditions upon the severance of a reversionary estate in land. The section provides for conditions and rights of re-entry, following a severance of the reversionary estate, to be apportioned to remain annexed to the severed parts of the reversionary estate. By s.140(2) a right of re-entry is defined to include a right to determine the lease by notice to quit, and by this subsection a notice served under this section to terminate the farm business tenancy on its term date is deemed to be a notice to quit.

This subsection ensures that the right of a tenant under s.140(2) of the Law of Property Act 1925 to choose to give a counter-notice to bring the entire tenancy to an end on the date upon which the original notice takes effect where he receives a notice to quit from the owner of one part of the severed reversionary estate, is preserved.

Length of notice to quit

6.—(1) Where a farm business tenancy is a tenancy from year to year, a notice to quit the holding or part of the holding shall (notwithstanding any provision to the contrary in the tenancy) be invalid unless—
   (a) it is in writing,
   (b) it is to take effect at the end of a year of the tenancy, and
   (c) it is given at least twelve months but less than twenty-four months before the date on which it is to take effect.

(2) Where, by virtue of section 5(1) of this Act, a farm business tenancy for a term of more than two years is to continue (as from the term date) as a tenancy from year to year, a notice to quit which complies with subsection (1) above and which is to take effect on the first anniversary of the term date shall not be invalid merely because it is given before the term date; and in this subsection "the term date" has the meaning given by section 5(2) of this Act.

(3) Subsection (1) above does not apply in relation to a counter-notice given by the tenant by virtue of subsection (2) of section 140 of the Law of Property Act 1925 (apportionment of conditions on severance of reversion).

DEFINITIONS
"farm business tenancy": s.1.
"given": s.36.
"holding": s.38(1).
"tenancy": s.38(1).
"tenant": s.38(1), (5).
"term date, the": s.5(2).

GENERAL NOTE
At common law, a notice to quit in respect of an annual periodic tenancy must be of at least six months' duration and take effect at the end of a year of the tenancy. This section overrides the common law position and provides for an extended notice period to be given by either landlord or tenant with no ability to contract out of its provisions. There is nothing in the Act, however, which interferes with the parties' ability to agree a surrender of the tenancy. There is no prescribed form for the notice and the common law rules must be complied with. Service of all notices and documents authorised or required to be served under this Act must be served in accordance with the provisions of s.36.

This section applies to tenancies which are annual periodic tenancies from the outset or those which were originally fixed-term tenancies of more than two years' duration which have not been terminated but which are continuing by reason of the provisions of s.5 above. There are no restrictions other than the length of notice requirement of this section and the service requirements in s.36, on the circumstances in which a landlord or a tenant can serve a notice to quit an annual periodic tenancy. This is a major change from the position under the 1986 Act where the landlord either had to have a ground set out in Sched. 3 to that Act which enabled him to serve an incontestable notice to quit or had to obtain the consent of the Agricultural Lands Tribunal to the operation of his notice to quit. Under this Act, once a tenant receives a notice to quit, provided that it complies with the common law rules as to validity, the length of notice requirements of this section and the service requirements of s.36, there is nothing that he can do to prevent it taking effect. Other periodic tenancies, weekly, monthly or quarterly, are not governed by the Act and the common law rules as to notice period will apply.

Subs. (2)
This subsection provides for the situation which falls between s.5 and s.6. Section 5 envisages the service of a notice to bring a fixed-term tenancy to an end on its term date; s.6(1) provides for the service of a notice to quit an annual periodic tenancy. A notice given in the final year of a fixed-term tenancy to expire on the first anniversary of the end of the fixed-term is neither a notice under s.5 or a notice under s.6 as the tenancy is not yet a tenancy from year to year. The validity of such a notice is preserved by this subsection.

Subs. (3)
The counter-notice referred to is that mentioned in the notes to s.5(3) above.

Notice required for exercise of option to terminate tenancy or resume possession of part

7.—(1) Where a farm business tenancy is a tenancy for a term of more than two years, any notice to quit the holding or part of the holding given in pursuance of any provision of the tenancy shall (notwithstanding any provision to the contrary in the tenancy) be invalid unless it is in writing and is given at least twelve months but less than twenty-four months before the date on which it is to take effect.

(2) Subsection (1) above does not apply in relation to a counter-notice given by the tenant by virtue of subsection (2) of section 140 of the Law of Property Act 1925 (apportionment of conditions on severance of reversion).

(3) Subsection (1) above does not apply to a tenancy which, by virtue of subsection (6) of section 149 of the Law of Property Act 1925 (lease for life or lives or for a term determinable with life or lives or on the marriage of the lessee), takes effect as such a term of years as is mentioned in that subsection.

DEFINITIONS
"farm business tenancy": s.1.
"given": s.36.
"holding": s.38(1).

"tenancy": s.38(1).
"tenant": s.38(1), (5).

GENERAL NOTE
This section controls break clauses in fixed-term tenancies of more than two years by imposing a statutory restriction on the length of notice to be given to operate a contractual option to break. As a result of this length of notice requirement, care should be taken in advising landlords who may need to be able to resume possession quickly, for example, for the purposes of redevelopment. It may be necessary (if practical in the particular circumstances of the case) to have a separate tenancy of any parts where this is likely to be an issue. Such a tenancy could either be a periodic (say, quarterly) farm business tenancy where only one period's notice would need to be given or, if the primary use of that part is not agricultural, a fixed-term tenancy under the Landlord and Tenant Act 1954, Pt. II, excluded by an agreement approved by the courts from the security of tenure provisions of that Act (see s.38 of the 1954 Act). Break clauses in such an excluded 1954 Act tenancy would not be subject to this length of notice requirement.

Contracting out of this provision is not permitted, although a consensual surrender between the parties is always a possibility. A break clause which allows for a shorter period of notice will probably not be void, but the notice requirements will be overridden by the statute.

Subs. (2)
The counter-notice referred to is that mentioned in the notes to s.5(3) above.

Subs. (3)
Section 149(6) of the Law of Property Act 1925 converts certain leases which are for life or determinable upon the tenant's marriage into 90-year terms determinable after death or marriage (as the case may be) by one month's notice expiring on a quarter day applicable to the tenancy or, if none, on a usual quarter day. The provisions of s.7 relating to break clauses do not apply to such leases and, therefore, they continue to be determinable by one quarter's notice as stated above.

Tenant's right to remove fixtures and buildings

Tenant's right to remove fixtures and buildings

8.—(1) Subject to the provision of this section—
(a) any fixture (of whatever description) affixed, whether for the purposes of agriculture or not, to the holding by the tenant under a farm business tenancy, and
(b) any building erected by him on the holding,
may be removed by the tenant at any time during the continuance of the tenancy or at any time after the termination of the tenancy when he remains in possession as tenant (whether or not under a new tenancy), and shall remain his property so long as he may remove it by virtue of this subsection.
(2) Subsection (1) above shall not apply—
(a) to a fixture affixed or a building erected in pursuance of some obligation,
(b) to a fixture affixed or a building erected instead of some fixture or building belonging to the landlord,
(c) to a fixture or building in respect of which the tenant has obtained compensation under section 16 of this Act or otherwise, or
(d) to a fixture or building in respect of which the landlord has given his consent under section 17 of this Act on condition that the tenant agrees not to remove it and which the tenant has agreed not to remove.
(3) In the removal of a fixture or building by virtue of subsection (1) above, the tenant shall not do any avoidable damage to the holding.
(4) Immediately after removing a fixture or building by virtue of subsection (1) above, the tenant shall make good all damage to the holding that is occasioned by the removal.
(5) This section applies to a fixture or building acquired by a tenant as it applies to a fixture or building affixed or erected by him.
(6) Except as provided by subsection (2)(d) above, this section has effect notwithstanding any agreement or custom to the contrary.

(7) No right to remove fixtures that subsists otherwise than by virtue of this section shall be exercisable by the tenant under a farm business tenancy.

DEFINITIONS
"agriculture": s.38(1).
"building": s.38(1).
"farm business tenancy": s.1.
"holding": s.38(1).
"landlord": s.38(1), (5).
"tenancy": s.38(1).
"tenant": s.38(1), (5).
"termination": s.38(1).

GENERAL NOTE
At common law, whether a particular object on land is a fixture or a chattel depends to an extent upon the degree of annexation of the object to the land, the ease with which it can be removed and, more importantly, whether the purpose for which it was originally attached was for the permanent improvement of the land or for some temporary purpose. It is beyond the scope of this commentary to go into detail (see Woodfall, *Landlord and Tenant*, Sweet & Maxwell, volume 1, paras. 13.131 *et seq*). Chattel assets can always be removed by a tenant at any time. Fixtures can be removed in certain circumstances although at common law they become attached to the land and the property of the freehold owner. The common law was relaxed in relation to trade fixtures (*Pooles case* (1703) 1 SALK 368) but this did not extend to general agricultural fixtures not of a specialised nature (see *Elwes v. Maw* [1802] 3 East 38).

Under s.10 of the 1986 Act, the tenant is given a right to remove certain fixtures provided that two conditions are fulfilled. The first is that the tenant must have paid all of the rent and complied with all of the terms and conditions of his tenancy. Should a landlord wish to take the point, compliance with this condition is extremely difficult in the context of, particularly, a tenant's repairing obligation. The second condition requires a notice to be served on the landlord stating the tenant's intention to remove the fixture. Where a tenant serves such a notice the landlord can elect to purchase the fixture and hence override the tenant's right to remove.

By this new section the tenant's position is greatly improved: there is no requirement that he must have complied with the terms and conditions of his tenancy and there is no obligation to serve a notice of intention to remove. The landlord cannot elect to purchase the fixture and override the tenant's right to remove although, as can be seen below (subs. (2)(d)), there are circumstances in which the landlord can indirectly force the tenant to treat the fixture as an improvement and take compensation instead of removing it at the end of the tenancy.

If the tenant chooses not to remove the fixture, it may in certain circumstances be treated as an improvement for which the tenant will be paid compensation (see the General Notes to Pt. III below).

Subs. (1)
The right to remove is not confined to general agricultural fixtures but extends to other trade fixtures or, indeed, to those affixed merely for domestic or ornamental purposes and to buildings provided, in all cases, that they were affixed by the tenant. The tenant may remove, without notice being given to the landlord, during the tenancy or after termination provided that he remains in possession "as a tenant". By s.38(5) the designations of landlord and tenant continue to apply until the conclusion of any proceedings under the Act relating to compensation, but that does not, of course, mean that the tenant remains in possession as tenant. The tenant will remain in possession as tenant if he is holding over or if he takes a new tenancy but if he fails to leave after the expiry of a notice to quit or after forfeiture has been effected, he will not be in possession as a tenant but as a trespasser.

For so long as the tenant has a right to remove in accordance with this section, the common law position as to title to the fixture is altered. At common law, regardless of any rights of removal, title to the fixture vests in the landlord at the moment that the fixture is affixed. Under this subsection, property remains in the tenant and is therefore vulnerable to any remedies or enforcement methods which can be taken against the tenant's property.

Subs. (2)
This subsection circumscribes the tenant's right to remove set out in subs. (1) above.

Paragraphs (a) and (b) also appear as restrictions in the 1986 Act. The obligation referred to in para. (a) may be an obligation imposed by the tenancy but may equally be a statutory obligation. In cases falling within para. (a), the tenant may nevertheless be entitled to compensation for the item as an improvement (see the General Notes to Pt. III below).

Paragraph (c) effectively presents the tenant, in some cases, with an option. Some improvements which would fall to be compensated under s.16 of the Act will fall within the definition of fixture. Under the 1986 Act, if a fixture could be compensated as an improvement, the tenant could not choose to remove it as a fixture at the end of the tenancy but had to rely on his compensation claim. Under this paragraph, it is only if the tenant has actually obtained compensation for the fixture as an improvement that he is prevented from removing it. At the end of his farm business tenancy, therefore, where the tenant has a right to compensation, subject to para. (d) below, he can choose to remove the improvement as a fixture. In many cases, this will be far more straightforward for a tenant than going through arbitration to obtain compensation where he cannot reach agreement as to the amount of compensation with the landlord. Care needs to be taken, however, to ensure that the time taken up in attempting to negotiate compensation does not put the tenant out of time for removing as a fixture.

Paragraph (d) refers to s.17. Section 17 sets out the conditions of eligibility for compensation for tenants' improvements. Before a tenant is entitled to compensation, he must have obtained the landlord's written consent to the improvement. That consent may be conditional provided that the condition imposed relates to the improvement itself. If a condition is that the tenant shall not be entitled to treat the improvement as a fixture, the tenant must rely on his right to compensation and may not rely on the right to remove set out in subs. (1) above.

Subss. (3) and (4)
There is a positive obligation not to do avoidable damage and an obligation to make good all damage whether avoidable or otherwise. The need for subs. (3) in addition to subs. (4) is to enable the landlord to obtain an injunction if necessary and not to have to get into a position of "suffering" the damage and seeking damages from a tenant who may not be in a position to pay.

Subs. (5)
This subsection may apply, for example, where the fixture was acquired from a previous tenant.

Subs. (6)
No contracting out of this section is permitted save insofar as a landlord, when giving written consent to an improvement, may impose a condition that the tenant may not remove the improvement as a fixture. This prohibition on contracting out can be compared with the uncertain position in connection with s.10 of the 1986 Act where it was at least arguable that the parties could contract out (see *Premier Dairies v. Garlick* [1920] 2 Ch. 17 although this was doubted in *Johnson v. Moreton* [1980] AC 37).

Whilst contracting out of the right to remove is not permitted, there is nothing to prevent the parties agreeing to a clause in the tenancy agreement saying that a tenant cannot erect fixtures or cannot erect fixtures of a particular type. However, if such a fixture would amount to an improvement under the definition contained in s.15 below such a clause could, it is submitted, be overridden by arbitration such as to allow the improvement to be made (see the General Note to s.19 below). If such a clause is overridden, it is not clear whether that entitles the tenant to take the improvement, put up after the arbitrator's consent, as a fixture or whether he is confined to his compensation claim. Whilst a landlord can, as a condition of giving consent to the improvement, prevent the tenant from removing it as a fixture, an arbitrator is not entitled to impose conditions upon his consent to an improvement (see s.19) and, should he choose to override a prohibition on the erection of fixtures, he might thereby be giving the tenant a free choice contrary to the terms of the tenancy agreement.

Subs. (7)
This subsection can be compared with the 1986 Act where the common law rules allowing removal of trade fixtures still apply in relation to a 1986 Act tenancy. However, s.10 of the 1986 Act was drafted much more tightly and there were far greater conditions upon the ability of a tenant to remove fixtures.

Part II

Rent Review Under Farm Business Tenancy

Introduction
In the initial Consultation Paper produced by MAFF and WOAD in February 1991, it was proposed that there should be no regulation at all in relation to rent or rent reviews in the new tenancies regime. By the time of the detailed proposals paper in September 1992, fallback provisions relating to the timing of rent reviews and to the basis for determining the new rent were

envisaged, but the emphasis was still very much on complete freedom of contract with the new legislation filling in where the parties had not turned their minds to it. What has ended up in the Act is a more complicated compromise between freedom of contract and regulation such that, in broad terms: (1) the initial rent set can be any amount agreed between the parties; (2) the parties can choose any basis for rent review that they wish; (3) if the basis chosen is one of the options set out in s.9, the parties have to abide by it and cannot opt back into the statutory open market basis; (4) if the basis chosen is not one of the options set out in s.9, it will operate only for so long as both parties are content for it to operate, but either party may opt back into the statutory basis on any of the rent reviews (s.12(b) below); (5) if the parties say nothing about the basis of the rent reviews in their agreement, the statutory open market formula set out in s.13 will apply; (6) in any case, the parties can choose the frequency of their rent reviews.

Application of Part II

9. This Part of this Act applies in relation to a farm business tenancy (notwithstanding any agreement to the contrary) unless the tenancy is created by an instrument which—
    (a) expressly states that the rent is not to be reviewed during the tenancy, or
    (b) provides that the rent is to be varied, at a specified time or times during the tenancy—
        (i) by or to a specified amount, or
        (ii) in accordance with a specified formula which does not preclude a reduction and which does not require or permit the exercise by any person of any judgment or discretion in relation to the determination of the rent of the holding,
but otherwise is to remain fixed.

DEFINITIONS
    "farm business tenancy": s.1.
    "holding": s.38(1).
    "tenancy": s.38(1).

GENERAL NOTE
    Section 9 provides for the statutory rent review mechanism to apply unless the parties have agreed to one of the other options set out in s.9. Save for these options, contracting out of the statutory mechanism is not permitted although arbitration on the statutory *formula* is a last resort in the absence of agreement or some form of mediation (see s.12(b)). The options in s.9 are designed to avoid the long express rent review clauses found in commercial leases. However, where the parties do agree a formula which is not permitted by the statute, recognition is given to this in s.12(b) (see below). Evidential problems of showing the rent review agreement between the parties have been avoided by only permitting these options where they are set out in a written contract of tenancy between the parties. With oral lettings, the statutory mechanisms will, therefore, apply regardless of any agreement between the parties.
    Option (a). This may be attractive in short-term agreements which would nevertheless be long enough for the statutory mechanism to trigger a rent review in the absence of an express clause, for example, a five-year agreement. There may also be circumstances where both parties wish to avoid risk and where the tenant is prepared to pay more by way of an initial rent to avoid a statutory rent review. As a result of s.10(4) below, it is possible for the parties to agree that there should be no reviews for the early part of the term and then to commence reviews thereafter either under one of the two options set out in s.9(b) below or under the statutory rent review mechanism. This is because it is possible for the parties to choose the frequency of their reviews.
    Option (b)(i). It is important if this option is chosen that the tenancy agreement makes it clear that, other than the stated increases or decreases, the rent is to remain fixed. If not, the parties' choice will not prevail over the statutory mechanism. The circumstances in which this option is chosen are likely to be special: neither landlord nor tenant in a long fixed-term lease are likely to want to second guess the market at the outset. It may be something which the parties will want if, for example, the length of the agreement is such that there will only be one review or the length of the agreement is such that they would have agreed no rent review but either (a) the tenant is

required to do works on the holding such that a lower than usual rent is negotiated for the early part of the term or (b) a higher than usual rent is negotiated by way of a rentalised premium for the early part of the term and the arrangement, in either case, is that an appropriate increase or decrease in the rent will take place part way through the term.

Option (b)(ii). The formula chosen under this option must be objective, require no judgement or discretion to be exercised and cannot be upwards only. Effectively, the formula must be one which leads to a mathematical and certain calculation with no room for dispute between the parties. Indexation of some kind is the most obvious example of such a formula although care would have to be taken in identifying the relevant index, particularly in a long-term lease where changes to indices are possible, and in choosing the index: movements in the Retail Prices Index, for example, may bear little relation to farm profitability or rental levels.

Initial concerns were expressed that where the compilation of an index itself required the exercise of judgement or discretion, for example in the choice of commodities to be included in the index, such exercise would fall foul of the limits of this option. It is submitted that the exercise of judgement or discretion in the compilation of the index does not present a problem: it is only in connection with the application of the index to the rent that pure objectivity is necessary. The section makes it clear that judgement or discretion is outlawed "... in relation to the determination of the rent of the holding".

Alternative formulations may be linked to commodity prices, whether one particular commodity or a basket of commodities, although careful drafting will be needed to ensure complete clarity as to the method of determination of the price, or to a turnover rent in particular circumstances.

As in relation to option (b)(i) above, this option, if chosen, must provide for the rent to remain fixed other than for changes in relation to the application of the formula. If not, the statutory rent review mechanisms will prevail, not as additional reviews but as the only method by which the rent can be reviewed as a result of the fact that the option will not be in accordance with those permitted under s.9 (subject to s.12(b) below).

Such are the strictures of the formula under s.9(b)(ii) that achieving the equivalent to a sitting tenant rent under s.12 of and Sched. 2 to the 1986 Act would be difficult if not impossible. This will create stumbling blocks for landlords and tenants who wish to effect deals, either by an express surrender and regrant, or by moving a 1986 Act protected tenant to another farm on the estate on or after September 1, 1995. The loss of security of tenure for such a tenant may, to an extent, be compensated by a long fixed term, but the inability of the landlord to guarantee 1986 Act level rents may be difficult to overcome.

In respect of the above options, the timing of reviews is a matter for agreement between the parties. As the rest of this Part of the Act only applies where one of the options is not chosen, the parties will also have to be specific on trigger mechanisms for instigating reviews if that is what they want.

Notice requiring statutory rent review

10.—(1) The landlord or tenant under a farm business tenancy in relation to which this Part of this Act applies may by notice in writing given to the other (in this Part of this Act referred to as a "statutory review notice") require that the rent to be payable in respect of the holding as from the review date shall be referred to arbitration in accordance with this Act.

(2) In this Part of this Act "the review date", in relation to a statutory review notice, means a date which—
  (a) is specified in the notice, and
  (b) complies with subsections (3) to (6) below.

(3) The review date must be at least twelve months but less than twenty-four months after the day on which the statutory review notice is given.

(4) If the parties have agreed in writing that the rent is to be, or may be, varied as from a specified date or dates, or at specified intervals, the review date must be a date as from which the rent could be varied under the agreement.

(5) If the parties have agreed in writing that the review date for the purposes of this Part of this Act is to be a specified date or dates, the review date must be that date or one of those dates.

(6) If the parties have not agreed as mentioned in subsection (4) or (5) above, the review date—
  (a) must be an anniversary of the beginning of the tenancy or, where the landlord and the tenant have agreed in writing that the review date for

the purposes of this Act is to be some other day of the year, that day of the year, and

(b) must not fall before the end of the period of three years beginning with the latest of any of the following dates—
   (i) the beginning of the tenancy,
   (ii) any date as from which there took effect a previous direction of an arbitrator as to the amount of the rent,
   (iii) any date as from which there took effect a previous determination as to the amount of the rent made, otherwise than as arbitrator, by a person appointed under an agreement between the landlord and the tenant, and
   (iv) any date as from which there took effect a previous agreement in writing between the landlord and the tenant, entered into since the grant of the tenancy, as to the amount of the rent.

DEFINITIONS
"beginning of the tenancy": s.38(4).
"farm business tenancy": s.1.
"given": s.36.
"holding": s.38(1).
"landlord": s.38(1), (5).
"review date, the": subs. (2), s.14.
"Statutory review notice": subs. (1), s.14.
"tenancy": s.38(1).
"tenant": s.38(1), (5).

GENERAL NOTE
This section only applies if the parties have not agreed one of the options in s.9 above. It does three things: (1) sets out the mechanism by which a rent review is triggered; (2) provides that the review is by way of arbitration; and (3) enables the review date to be determined.

Either landlord or tenant can instigate the review.

The notice is similar to that under s.12 of the 1986 Act. Under that Act it has been held that the notice, once given, cannot unilaterally be withdrawn by the giver: if the other party wishes the rent review to go ahead then it must go ahead. The notice is a trigger for the statutory procedure (see *Buckinghamshire County Council v. Gordon* (1986) 279 E.G. 853). The trigger notice has no prescribed form but must be in writing and it must specify the review date. See s.36 for service methods.

The review date is defined in such a way that the parties can choose the frequency of their reviews and/or the actual date in any particular year from which a review will take effect even if they are content to rely on the rent being reviewed in accordance with the s.13 open market formula and have not chosen their own basis for review. If the parties do not choose the frequency of reviews, basically there is a three-yearly rent review cycle.

Subss. (4) to (6)

It is these subsections which effectively define the review date. The parties can agree: (1) when reviews are to commence; and/or (2) the intervals between reviews; and/or (3) the actual review date itself.

Any arrangement must be in writing. If it is not agreed in writing, the review date will be as defined in subs. (6). Unlike the options in s.9, the agreement in writing defining the review date does not need to be in a written tenancy agreement and these options are therefore available where the letting is oral or where the parties reach a separate agreement. In the absence of agreement on the actual date of the review, the review date will be the anniversary of the date upon which the tenant is entitled to go into possession (which is the definition of the beginning of the tenancy – see s.38) and not necessarily the anniversary of the date of the tenancy agreement or the commencement of the term itself. This will apply where the parties have made no provision or where they have made provision for the intervals between reviews but have not specified the date itself.

The parties may, of course, agree a review date but be content with the statutory three-yearly intervals and may wish to do so where, for some reason, the beginning of the tenancy (when the tenant is entitled to go into possession) does not coincide with the usual rent day.

The net effect of subs. (6) is that the earliest date possible for the first rent review under the statutory fall back provisions is three years after the date upon which the tenant is entitled to go into possession. Thereafter rent reviews will take place at three-yearly intervals from the date of the last review, regardless of whether that review was by arbitration or by some other person appointed by the parties (s.12(b)) or by agreement and, more importantly, whether or not the last review was comprehensive or not. The Act does not deal with the problem highlighted in *Mann v. Gardner* (1990) 61 P. & C.R. 1 where the Court of Appeal held that a rent reduction of £100 on surrender of the farmhouse out of a total rent of over £21,000 was sufficient to prevent a comprehensive review of the rent for three years thereafter. Other minor alterations in rent which would not have triggered the review period under the 1986 Act will trigger the period under this Act; there is no equivalent to the provisions in Sched. 2, para. 4(2) to the 1986 Act which allow certain changes in the rent to be disregarded for the purposes of ascertaining whether or not the review period has been triggered. These small alterations are as a result of specific provisions in the 1986 Act which are not replicated in the 1995 Act.

Three-yearly reviews are familiar from the 1986 Act. However, there is a difference. Under the 1986 Act, the three years from the date of an agreement between the parties as to the rent only begins to run if the agreement was to increase or decrease the rent not if there was an agreement that the rent should stay at the same level (although a direction of the arbitrator that the rent should remain unchanged does start time running). Under this Act, any agreement as to the amount of the rent between the parties will start time running.

Review date where new tenancy of severed part of reversion

11.—(1) This section applies in any case where a farm business tenancy ("the new tenancy") arises between—
   (a) a person who immediately before the date of the beginning of the tenancy was entitled to a severed part of the reversionary estate in the land comprised in a farm business tenancy ("the original tenancy") in which the land to which the new tenancy relates was then comprised, and
   (b) the person who immediately before that date was the tenant under the original tenancy,
and the rent payable under the new tenancy at its beginning represents merely the appropriate portion of the rent payable under the original tenancy immediately before the beginning of the new tenancy.

(2) In any case where this section applies—
   (a) references to the beginning of the tenancy in subsection (6) of section 10 of this Act shall be taken to be references to the beginning of the original tenancy, and
   (b) references to rent in that subsection shall be taken to be references to the rent payable under the original tenancy,
until the first occasion following the beginning of the new tenancy on which any such direction, determination or agreement with respect to the rent of the new holding as is mentioned in that subsection takes effect.

DEFINITIONS
   "beginning of the tenancy": s.38(4).
   "farm business tenancy": s.1.
   "holding": s.38(1).
   "new tenancy, the": subs. (1).
   "original tenancy, the": subs. (1)(a).
   "tenancy": s.38(1).
   "tenant": s.38(1), (5).

GENERAL NOTE
   Since the Court of Appeal decision in *Jelley v. Buckman* [1974] Q.B. 488 (see also the later county court decisions of *Styles v. Farrow* [1977] 241 E.G. 623 and *Greenway v. Tempest* (1983) unreported) it has been clear that a severance of the landlord's reversionary estate does not bring about separate tenancies on those severed parts: the original tenancy continues, simply now with more than one landlord. Section 140 of the Law of Property Act 1925 then provides that each landlord may enforce covenants and may exercise rights of re-entry and notices to quit independently of one another. As a result of the decision in *Jelley v. Buckman*, there are only limited circumstances in which this section could bite. The first is where, following the severance of the reversion, there is an express agreement with the tenant to enter into new tenancies of each severed part. The second is where the tenant is a party to the deed of severance and agrees the apportionment of the rents between the severed parts where new tenancies of each part may

be implied. The section will, however, only bite where the rent under each new tenancy is merely the appropriate portion of the original rent. There is nothing to prevent the parties agreeing some other rent or basis for apportioning the rent at that stage.

There may actually be good reason for landlords on a severed reversion to have new tenancies. Whilst it is clear that s.140 of the Law of Property Act 1925 (see above) enables each landlord to enforce covenants and serve notices to quit, rent reviews are more difficult. In *Styles v. Farrow* (above) it was held that following a severance of the reversion, the rent could only be reviewed as to the whole holding and that separate demands for rent from each reversioner were invalid: s.140 of the Law of Property Act 1925 does not extend to apportion statutory procedures as opposed to contractual arrangements between the parties and, therefore, all landlords of a severed estate must join in the service of a statutory review notice. Where, however, the parties have chosen a s.9 option and there is a contractual trigger notice, s.140 may operate to apportion that contractual right to each and every reversioner.

Where s.11 bites, it continues the rent review timetable through to the new tenancies to ensure a continuation of the three-year cycle. If, for example, the rent was reviewed (whether by arbitration, determination or agreement) 12 months prior to the beginning of the new tenancies, the first rent review that can take place under the new tenancies is two years later and not three years later.

Appointment of arbitrator

12. Where a statutory review notice has been given in relation to a farm business tenancy, but—
   (a) no arbitrator has been appointed under an agreement made since the notice was given, and
   (b) no person has been appointed under such an agreement to determine the question of the rent (otherwise than as arbitrator) on a basis agreed by the parties,
either party may, at any time during the period of six months ending with the review date, apply to the President of the Royal Institution of Chartered Surveyors (in this Act referred to as "the RICS") for the appointment of an arbitrator by him.

DEFINITIONS
   "farm business tenancy": s.1.
   "given": s.36.
   "review date, the": ss.10(2), 14.
   "RICS, the": ss.12, 37(1).
   "statutory review notice": ss.10(1), 14.

GENERAL NOTE
   Section 12 must be read together with s.30 which sets out general provisions dealing with the appointment of an arbitrator under the Act. Section 28, which sets out the timetable for general arbitrations, does not apply to a rent arbitration (see s.28(5)), although any arbitration in respect of the rent will still be under the Arbitration Acts 1950 to 1979. If no arbitrator is appointed by agreement, application can be made by either party to the President of the RICS in the final six-month run up to the review date. The application need not be made by the person who served the statutory review notice.

   This section makes it clear that the ability of either party to apply to the President ends on the review date and, therefore, if agreement has not been reached as the review date nears, a protective application should be made by the party wanting the rent to be reviewed to preserve the position.

   If the parties agree a basis for reviewing the rent which is either: (a) one of the options set out in s.9 but not contained in a written tenancy agreement; or (b) some other basis for reviewing the rent, whether contained in a written tenancy agreement or not, s.12(b) provides a mechanism for allowing the rent to be determined on that basis with the assistance of a third party, provided that (a) that third party is not appointed as arbitrator and (b) the parties both still want the rent to be determined on that basis. The difference between the circumstances set out in (a) and (b) above and the s.9 options is that, in this case, neither party can be forced to follow through the agreed basis but can opt back into arbitration and the statutory basis for review.

   If a s.10 notice is given the parties may agree any basis for reviewing the rent or agree to follow through such a basis as set out in their tenancy agreement and appoint their third party to determine the rent on that agreed basis. Each party is, by the appointment of that third party (likely in most cases to be an expert), then disabled from opting for arbitration because s.12 does not then permit an application to be made to the President of the RICS. If the parties do not agree the appointment of their expert or their third party, application can be made for an arbitrator to be appointed but the arbitrator can only review the rent in accordance with s.13 and not in accordance with the agreement between the parties. Each party will make an assessment as to whether the open market formula or the chosen basis is likely to provide a better result for them and will act accordingly.

The upshot of the interrelationship between s.9 and s.12(b) is that the parties can choose any rent review mechanism they wish but they will only be bound to follow a rent review formula which is not the statutory formula or one of the options set out in s.9 if they have reached the point of agreeing the appointment of their expert. Either party can opt out up to that point. This fits with the interrelationship between s.28 and s.29 for other arbitrations and alternative dispute resolution mechanisms under the Act.

Mention was made earlier of the difficulty in re-organising estates after September 1, 1995 for tenancies with existing 1986 Act security of tenure. Whilst if a tenant is moved to another farm and put on to a farm business tenancy a long fixed-term tenancy can, to an extent, compensate for the lack of security of tenure of the 1986 Act, a sticking point may well be the inability of the parties to guarantee a s.12, Sched. 2 rent equivalent for the new tenancy. It is clear that the parties can, in their tenancy agreement, put in that the rent should be reviewed in accordance with s.12 of and Sched. 2 to the 1986 Act by an expert but it is equally clear that, on any particular review, the landlord or the tenant can revert to the open market rent formula in s.13 below by simply refusing to agree the appointment of the necessary expert to deal with the review on the basis of s.12 and Sched. 2.

Amount of rent

13.—(1) On any reference made in pursuance of a statutory review notice, the arbitrator shall determine the rent properly payable in respect of the holding at the review date and accordingly shall, with effect from that date, increase or reduce the rent previously payable or direct that it shall continue unchanged.

(2) For the purposes of subsection (1) above, the rent properly payable in respect of a holding is the rent at which the holding might reasonably be expected to be let on the open market by a willing landlord to a willing tenant, taking into account (subject to subsections (3) and (4) below) all relevant factors, including (in every case) the terms of the tenancy (including those which are relevant for the purposes of section 10(4) to (6) of this Act, but not those relating to the criteria by reference to which any new rent is to be determined).

(3) The arbitrator shall disregard any increase in the rental value of the holding which is due to tenant's improvements other than—
   (a) any tenant's improvement provided under an obligation which was imposed on the tenant by the terms of his tenancy or any previous tenancy and which arose on or before the beginning of the tenancy in question,
   (b) any tenant's improvement to the extent that any allowance or benefit has been made or given by the landlord in consideration of its provision, and
   (c) any tenant's improvement to the extent that the tenant has received any compensation from the landlord in respect of it.

(4) The arbitrator—
   (a) shall disregard any effect on the rent of the fact that the tenant who is a party to the arbitration is in occupation of the holding, and
   (b) shall not fix the rent at a lower amount by reason of any dilapidation or deterioration of, or damage to, buildings or land caused or permitted by the tenant.

(5) In this section "tenant's improvement", and references to the provision of such an improvement, have the meaning given by section 15 of this Act.

DEFINITIONS
   "beginning of the tenancy": s.38(4).
   "holding": s.38(1).

"landlord": s.38(1), (5).
"rent properly payable": subs. (2).
"review date, the": ss.10(2), 14.
"statutory review notice": ss.10(1), 14.
"tenancy": s.38(1).
"tenant": s.38(1), (5).
"tenant's improvement": subs. (5), s.15.

GENERAL NOTE

This section sets out the arbitrator's task on an appointment following a statutory review notice. The arbitrator must determine the rent properly payable as at the review date (not the date of reference to arbitration as under the 1986 Act) and the section sets out how that figure is to be found.

Subs. (2)

In a departure from the sitting tenant approach of s.12 of and Sched. 2 to the 1986 Act, the rent payable under the statutory review mechanism is the open market rent with minimal disregards. In comparison with the 1986 Act, the arbitrator no longer has to consider the productive capacity or related earning capacity of the holding and the arbitrator does not have to assume that the parties are "prudent" but simply "willing".

The arbitrator must look at all of the relevant factors but the open market formula is based on a hypothetical landlord and a hypothetical tenant. This is confirmed by subs. (4)(a) where the fact that the tenant is in occupation of the holding is to be disregarded. Circumstances personal to the parties can therefore be ignored. The terms of the tenancy, including the length of the unexpired term, the user covenants, the repairing obligations *etc.* and expressly the rent review intervals and dates (but not any references in the tenancy to a rent review formula) must be considered.

Tenants' improvements, insofar as they would otherwise lead to a rental increase, are to be ignored save in those circumstances set out in s.13(3)(a)–(c). Grant aid provided to either landlord or tenant to assist in the provision of the improvement does not feature so that where grant aid has been provided to a tenant the improvement may still be disregarded, although the definition of tenant's improvement must be considered as it is only those improvements which are provided by the tenant by his own effort or wholly or partly at his own expense which are tenants' improvements within the meaning of the Act (see s.15 below).

Scarcity can be taken into account as can marriage value save insofar as it arises from the tenant's occupation of the holding which is a specific disregard.

Subs. (3)(a)

Subsection (3)(a) disposes of one problem which arose from a similar provision in the 1986 Act. Under the 1986 Act tenants' improvements were brought into account if provided under an obligation imposed on the tenant by the terms of his contract of tenancy. It was unclear whether an improvement resulting from the landlord's conditional consent during the tenancy, which led to a variation of the tenancy agreement, would be caught. Under this subsection it is clear that such improvements would be taken into account on a rent review and that the disregard only relates to improvements provided pursuant to an obligation which arose before or at the time of the tenant becoming entitled to possession under the terms of his tenancy agreement.

Under the 1986 Act, the arbitrator was directed to consider (see Sched. 2 to that Act) the level of rents on comparable holdings, and other arbitrators' awards could be submitted in evidence. Under the 1995 Act, other arbitrators' awards may be inadmissible (see *Land Securities v. Westminster City Council* [1993] 1 W.L.R. 286).

Subs. (3)(b)

An improvement may still be a tenant's improvement within the meaning of s.15 if the landlord has assisted in paying for it, whether directly or whether by a rent reduction or other indirect benefit.

Subs. (3)(c)

This would cover, for example, an improvement provided by the tenant under a previous tenancy in respect of which he has received compensation.

Interpretation of Part II

14. In this Part of this Act, unless the context otherwise requires—
"the review date", in relation to a statutory review notice, has the meaning given by section 10(2) of this Act;

"statutory review notice" has the meaning given by section 10(1) of this Act.

DEFINITIONS
"the review date": ss.10(2), 14.
"statutory review notice": ss.10(1), 14.

PART III

COMPENSATION ON TERMINATION OF FARM BUSINESS TENANCY

Tenant's entitlement to compensation

INTRODUCTION
Under the 1995 Act the only compensation payable on termination is for tenants' improvements as defined by s.15. Any other matter, whether for a landlord or for a tenant, will be a matter of contract between the parties. This Part sets out a new and wide-ranging definition of tenants' improvements and a discrete regime for establishing eligibility and the amount of compensation to be paid.

Meaning of "tenant's improvement"

15. For the purposes of this Part of this Act a "tenant's improvement", in relation to any farm business tenancy, means—
  (a) any physical improvement which is made on the holding by the tenant by his own effort or wholly or partly at his own expense, or
  (b) any intangible advantage which—
    (i) is obtained for the holding by the tenant by his own effort or wholly or partly at his own expense, and
    (ii) becomes attached to the holding,
and references to the provision of a tenant's improvement are references to the making by the tenant of any physical improvement falling within paragraph (a) above or the obtaining by the tenant of any intangible advantage falling within paragraph (b) above.

DEFINITIONS
"farm business tenancy": s.1.
"holding": s.38(1).
"tenant": s.38(1), (5).
"tenant's improvement": s.15.

GENERAL NOTE
This section, and its wide definition of tenants' improvements, can be compared with the extensive lists of different types of improvement all with their own rules under the 1986 Act. The idea of the new definition is to avoid lists which can go out of date and need amendment and to choose a definition which can grow with the changes in the industry.

There are basically two classes of improvement: physical improvement and intangible advantages. As will be seen below, for some purposes, those two classes have been further broken down. Certain types of physical improvements, called "routine improvements" (see below) have different rules so far as references to arbitration are concerned and certain types of intangible advantages, namely planning permissions, have their own rules for eligibility and amount of compensation payable. Physical improvements may now also be fixtures (see s.8 above). This can be compared with the position under the 1986 Act where if a particular item was capable of attracting compensation as an improvement, it could not be treated as a fixture by the tenant. Here, it is only where compensation has actually been paid to the tenant that that restriction applies.

The Act allows compensation not only where a tenant has paid in whole or in part for the provision of the improvement but also where the improvement is made by the tenant's own effort, for example, his own labour. This is a change from the position under the 1986 Act and may be wide enough to cover not only the position where no costs other than the tenant's own efforts are involved but also to compensate the tenant where the landlord has provided the materials and the tenant the labour. As we shall see, the compensation payable to a tenant may be reduced where the landlord has paid for the provision of the improvement, but only if there is an agreement in writing to that effect (see s.20(2)) and there may be cases therefore where the landlord has paid for the materials and the tenant has provided the labour but where the tenant reaps the entire benefit on termination.

During the passage of the Bill, attempts were made to try to ensure that express provisions were included to provide for compensation to a tenant on quitting for tenant-right matters. The Bill always provided for compensation on quitting for physical improvements and there is no doubt that certain tenant-right matters, such as ploughing, liming, fertilizing, *etc.* would fall within that definition. During the final stages of the Bill in the House of Commons, however, a sub-category of physical improvements, called routine improvements was introduced. Routine improvements are defined in s.19(10) below and cover some, although not all, of those matters which would be tenant-right matters in the 1986 Act (see s.65 of and Sched. 8 to the 1986 Act). The only purpose of the subcategory (see s.19(2)) is to enable the Act to provide that the tenant may, in the case of a routine improvement but not otherwise, go to arbitration following a landlord's refusal to consent to the improvement, even where the tenant has begun the improvement in question.

Intangible advantages are a new concept and recognise that value may be added to a holding other than by the provision of physical improvements. It is only those intangible advantages which become attached to the holding and which remain on the holding at the termination of the tenancy in respect of which compensation may be paid.

Specifically provided for in the Act are planning permissions where the specified physical improvement has not been completed or the change of use effected. It is clear that milk quota is also intended to be compensated for as an intangible advantage and most would accept that, since the decision of Chadwick J in *Faulks v. Faulks* [1992] 1 E.G.L.R. 9, milk quota is attached to the holding despite the fact that in certain very limited circumstances a transfer of milk quota can now take place without a transfer of land (see reg. 13 of the Dairy Produce Quotas Regulations 1994 (S.I. 1994 No. 672)). Other quotas, subsidies and premiums payable are, at the moment, personal rights of the farmer but there is a certain amount of pressure to alter this position and to attach other subsidy payments to the land.

In addition to planning permissions and quotas, it is possible that goodwill and licences or designations, particularly of an environmental nature, may well fall to be treated as intangible advantages.

Tenant's right to compensation for tenant's improvement

16.—(1) The tenant under a farm business tenancy shall, subject to the provisions of this Part of this Act, be entitled on the termination of the tenancy, on quitting the holding, to obtain from his landlord compensation in respect of any tenant's improvement.

(2) A tenant shall not be entitled to compensation under this section in respect of—
  (a) any physical improvement which is removed from the holding, or
  (b) any intangible advantage which does not remain attached to the holding.

(3) Section 13 of, and Schedule 1 to, the Agriculture Act 1986 (compensation to outgoing tenants for milk quota) shall not apply in relation to a farm business tenancy.

DEFINITIONS
  "farm business tenancy": s.1.
  "holding": s.38(1).
  "landlord": s.38(1), (5).
  "tenant": s.38(1), (5).
  "tenant's improvement": s.15.
  "termination of the tenancy": s.38(1).

GENERAL NOTE
  Subsections (1) and (2) set out the entitlement to compensation and make it clear (if it were not already) that it is effectively only where the benefit of the improvement remains with the holding that the tenant is entitled to compensation. Furthermore, two things have to coincide

before the tenant is entitled to compensation; (1) the tenancy must be terminated, and (2) the tenant must be quitting the holding. If the tenant remains on the holding, for example, under a new tenancy, there is no *entitlement* at the termination of the earlier tenancy to compensation. Reference should be made to the General Note to s.23 below.

Subsection (3) ensures that the compensation provisions for outgoing tenants in relation to milk quotas set out in the Agriculture Act 1986 (c. 49) cannot apply to farm business tenancies and that the amount of compensation is by reference to s.20 of this Act. This is designed to prevent a tenant from claiming twice. Any tenant moving from a 1986 Act tenancy carrying with it the right to compensation for milk quota in accordance with the provisions of the Agriculture Act 1986 is, therefore, advised to make his claim for compensation before moving, and to reach agreement as to the amount of compensation due under the 1986 Act to fix entitlement at the termination of the 1986 tenancy. A tenant may, of course, be prepared for the landlord to delay actual payment.

Conditions of eligibility

Consent of landlord as condition of compensation for tenant's improvement

17.—(1) A tenant shall not be entitled to compensation under section 16 of this Act in respect of any tenant's improvement unless the landlord has given his consent in writing to the provision of the tenant's improvement.

(2) Any such consent may be given in the instrument creating the tenancy or elsewhere.

(3) Any such consent may be given either unconditionally or on condition that the tenant agrees to a specified variation in the terms of the tenancy.

(4) The variation referred to in subsection (3) above must be related to the tenant's improvement in question.

(5) This section does not apply in any case where the tenant's improvement consists of planning permission.

DEFINITIONS
"given": s.36.
"landlord": s.38(1), (5).
"planning permission": s.27 and the Town and Country Planning Act 1990, s.336(1).
"tenancy": s.38(1).
"tenant": s.38(1), (5).
"tenant's improvement": s.15.

GENERAL NOTE

Subs. (1)
Consent may be given after the improvement has been effected but the tenant runs a risk: if consent is refused then, save in the case of routine improvements (see s.19(2), (8) and (10) below), the tenant cannot then go to arbitration to obtain the approval of the arbitrator to the provision of the improvement. The requirement for consent is particularly important in the context of milk quota where the tenant may inadvertently transfer milk quota off another holding which he occupies (whether as owner-occupier or as tenant) on to the farm business tenancy land simply by using the land for milk production (within the wide meaning of *Puncknowle Farms v. Kane* [1985] 3 All E.R. 790). The moment that the tenant begins to use the farm business tenancy land in connection with his dairy business the improvement has begun, as the milk quota will begin to attach to the farm business tenancy land and it is then too late to go to arbitration should the landlord subsequently refuse consent. Careful drafting of tenancies for tenants or owners of dairy enterprises taking additional land on farm business tenancies is required to avoid this problem.

Subs. (2)
It is likely that consent will be given in the tenancy agreement to routine improvements. If consent is not given directly, but the tenant is placed under a positive obligation to effect the improvement, such an obligation will imply the consent of the landlord. For example, if a tenant is obliged to farm to a certain standard and husbandry rules are laid down in the tenancy agreement, the consent to routine improvements to farm to that standard will have been impliedly given.

Subss. (3) and (4)
Examples of permitted variations will be an alteration to the repairing covenant or (and this can be seen by s.8(2)(d)) a condition that the tenant will treat the improvement as an improvement and not remove it as a fixture on the termination of the tenancy.

Conditions in relation to compensation for planning permission

18.—(1) A tenant shall not be entitled to compensation under section 16 of this Act in respect of a tenant's improvement which consists of planning permission unless—
   (a) the landlord has given his consent in writing to the making of the application for planning permission,
   (b) that consent is expressed to be given for the purpose—
      (i) of enabling a specified physical improvement falling within paragraph (a) of section 15 of this Act lawfully to be provided by the tenant, or
      (ii) of enabling the tenant lawfully to effect a specified change of use, and
   (c) on the termination of the tenancy, the specified physical improvement has not been completed or the specified change of use has not been effected.
(2) Any such consent may be given either unconditionally or on condition that the tenant agrees to a specified variation in the terms of the tenancy.
(3) The variation referred to in subsection (2) above must be related to the physical improvement or change of use in question.

DEFINITIONS
   "given": s.36.
   "landlord": s.38(1), (5).
   "planning permission": s.27 and the Town and Country Planning Act 1990, s.336(1).
   "tenancy": s.38(1).
   "tenant": s.38(1), (5).
   "tenant's improvement": s.15.
   "termination of the tenancy": s.38(1).

GENERAL NOTE

Subs. (1)
This section sets out a separate, although similar, set of conditions of eligibility for planning permissions than for other improvements. Planning permissions are a species of intangible advantage. Compensation will only be available for planning permissions if the landlord's written consent has been obtained: there is no fall back of arbitration (see s.19(1)(a)).

The regime for compensating for planning permissions is to ensure that a tenant who, with consent, has applied for planning permission as a first stage of effecting a physical improvement or change of use (see subs. (1)(b)) but who has not completed the improvement or effected the change of use (see subs. (1)(c)) on the termination of the tenancy, is compensated. The compensation provisions are most likely to be relevant where a landlord forfeits the lease or exercises a break clause or where the tenant is on a continuing annual periodic tenancy and the landlord serves 12 months notice to quit: in other words, in circumstances where, from the tenant's point of view, the tenancy unexpectedly comes to an end before he has implemented the permission.

If planning permission is likely to improve the value of the holding (see s.20 below) and particularly if the application is going to cost the tenant a substantial amount of money, it would be wise to seek the landlord's consent specifically to the application for planning permission itself as well as to the improvement to cover the unexpected termination situation. If the landlord refuses consent to the planning permission and to the improvements, the tenant can still apply for planning permission and can go to arbitration in respect of the improvement (see below) but the planning permission would not be compensatable in those circumstances if the tenancy were to end before the physical improvement or change of use had been effected.

Of course, the landlord can himself apply for planning permission but it will be at his cost. Whether he does so or gives the tenant permission to do so will depend on the circumstances, including the costs of the planning permission against the likely cost of compensating the tenant for an unimplemented permission and the landlord's ability to terminate the lease and the likelihood that he will do so.

Subss. (2) and (3)
The conditions which may be imposed must relate to the physical improvement or change of use and not to the planning permission itself.

Reference to arbitration of refusal or failure to give consent or of condition attached to consent

19.—(1) Where, in relation to any tenant's improvement, the tenant under a farm business tenancy is aggrieved by—
  (a) the refusal of his landlord to give his consent under section 17(1) of this Act,
  (b) the failure of his landlord to give such consent within two months of a written request by the tenant for such consent, or
  (c) any variation in the terms of the tenancy required by the landlord as a condition of giving such consent,
the tenant may by notice in writing given to the landlord demand that the question shall be referred to arbitration under this section; but this subsection has effect subject to subsections (2) and (3) below.

(2) No notice under subsection (1) above may be given in relation to any tenant's improvement which the tenant has already provided or begun to provide, unless that improvement is a routine improvement.

(3) No notice under subsection (1) above may be given—
  (a) in a case falling within paragraph (a) or (c) of that subsection, after the end of the period of two months beginning with the day on which notice of the refusal or variation referred to in that paragraph was given to the tenant, or
  (b) in a case falling within paragraph (b) of that subsection, after the end of the period of four months beginning with the day on which the written request referred to in that paragraph was given to the landlord.

(4) Where the tenant has given notice under subsection (1) above but no arbitrator has been appointed under an agreement made since the notice was given, the tenant or the landlord may apply to the President of the RICS, subject to subsection (9) below, for the appointment of an arbitrator by him.

(5) The arbitrator shall consider whether, having regard to the terms of the tenancy and any other relevant circumstances (including the circumstances of the tenant and the landlord), it is reasonable for the tenant to provide the tenant's improvement.

(6) Subject to subsection (9) below, the arbitrator may unconditionally approve the provision of the tenant's improvement or may withhold his approval, but may not give his approval subject to any condition or vary any condition required by the landlord under section 17(3) of this Act.

(7) If the arbitrator gives his approval, that approval shall have effect for the purposes of this Part of this Act and for the purposes of the terms of the farm business tenancy as if it were the consent of the landlord.

(8) In a case falling within subsection (1)(c) above, the withholding by the arbitrator of his approval shall not affect the validity of the landlord's consent or of the condition subject to which it was given.

(9) Where, at any time after giving a notice under subsection (1) above in relation to any tenant's improvement which is not a routine improvement, the tenant begins to provide the improvement—
  (a) no application may be made under subsection (4) above after that time,

(b) where such an application has been made but no arbitrator has been appointed before that time, the application shall be ineffective, and
(c) no award may be made by virtue of subsection (6) above after that time except as to the costs of the reference and award in a case where the arbitrator was appointed before that time.
(10) For the purposes of this section—
"fixed equipment" includes any building or structure affixed to land and any works constructed on, in, over or under land, and also includes anything grown on land for a purpose other than use after severance from the land, consumption of the thing grown or its produce, or amenity;
"routine improvement", in relation to a farm business tenancy, means any tenant's improvement which—
(a) is a physical improvement made in the normal course of farming the holding or any part of the holding, and
(b) does not consist of fixed equipment or an improvement to fixed equipment,
but does not include any improvement whose provision is prohibited by the terms of the tenancy.

DEFINITIONS
"farm business tenancy": s.1.
"fixed equipment": subs. (10).
"given": s.36.
"landlord": s.38(1), (5).
"RICS": s.38(1).
"routine improvement": subs. (10).
"tenancy": s.38(1).
"tenant": s.38(1), (5).
"tenant's improvement": s.15.

GENERAL NOTE

Subs. (1)
The ability of the tenant to refer the matter of consent to an improvement to arbitration, is confined to the refusal to give consent under s.17. Section 17 does not apply to planning permissions and therefore it is not possible to apply to an arbitrator to override a landlord's refusal to grant consent to obtain planning permission. This means that where a landlord refuses consent to planning permission and to the improvement and the tenant obtains the consent of the arbitrator to the improvement, he must run the risk of obtaining planning permission and it not being implemented by the time of the termination of the tenancy.
In order to start time running against a landlord who does not respond, the tenant must make his request in writing (see s.19(1)(b)). In all cases, therefore, in order to save time, all requests for consent should be made in writing. The tenant can apply not only following a refusal or no response from the landlord but also if he feels "aggrieved" by the conditions which the landlord wishes to impose. If the tenant refers the matter to arbitration for this reason, he obviously runs the risk of a refusal of approval. However, that refusal would not put him in any worse a position (see subss. (7) and (8) below).
Section 28 of this Act does not apply to arbitrations under this section (s.28(5)) although s.30 does and s.29 (alternative dispute resolution) is not available as it only disapplies s.28 when s.28 would otherwise apply.
The arbitration is under the Arbitration Acts 1950 to 1979.

Subs. (2)
See the General Note to s.17(1) above and the commentary to subs. (9) below.

Subs. (3)(b)
In effect the landlord is deemed to have refused after two months and this timetable brings it into line with (3)(a) above. The tenant cannot let matters drift. Notice has to be given demanding

arbitration within a set time period. However, there is nothing to stop a tenant starting again by making a fresh request for consent.

Subs. (4)
There is no time-limit within which the tenant or the landlord must apply for the appointment of an arbitrator, although by subs. (9) below no application may be made after the tenant begins to provide the improvements. If an agreement between the parties as to the identity of the arbitrator is to oust the ability of the parties to apply to the President, it must be an agreement reached after notice is given under subs. (1) above. The identity of an arbitrator cannot, therefore, be by way of a standing agreement in the tenancy agreement itself.

Subs. (5)
The arbitrator is specifically directed to take into account the circumstances of the tenant and the landlord, and this may include their financial and personal circumstances. If the landlord is able to show that he will not be able to afford to pay compensation on the termination of the tenancy, this will be a consideration which the arbitrator may take into account. The arbitrator must also have regard to the terms of the tenancy and this will obviously include the length of the unexpired term and the ability of either party to bring it to an end before that date. It will also include any clause in the tenancy agreement prohibiting the improvement in question. However, it is submitted that the arbitrator is able to override such a prohibition, otherwise a refusal by the landlord in advance in the tenancy agreement could achieve what a refusal at the time of a request could not. It may be, however, that a refusal in the tenancy agreement would carry more weight: the tenant has known from the outset of the prohibition and that prohibition together with the other terms of the tenancy will have determined the rent payable.

Other considerations for the arbitrator will be the need for the improvement and the impact on the tenant's business if it is not provided, the likely life of the improvement against the unexpired term of the tenancy and whether the improvement is needed to comply with statutory obligations, for example relating to pollution control. The question for the arbitrator is whether it is reasonable for a tenant to provide the improvement not whether it is necessary.

Subss. (6)–(8)
The arbitrator is limited in what he can do: he can grant permission for the improvement or he can refuse it. He cannot impose conditions or, for example, state that the tenant may not remove the improvement as a fixture at the termination of the tenancy. If the arbitrator refuses consent and the tenant commenced arbitration because he did not like the landlord's qualified consent, subs. (8) ensures that the tenant is no worse off and can return and take the conditions imposed by that qualified consent. The landlord is not able to withdraw his consent once it has been given.

Subs. (9)
This subsection extends subs. (2) above effectively to prevent a tenant from obtaining compensation for an improvement which he begins to implement before the arbitration award has been made, by ensuring that he is not entitled to continue the arbitration process if he begins the improvement.

Subs. (10)
Routine improvements are essentially certain tenant-right matters, being physical improvements made in the normal course of farming. A wide definition, avoiding lists, has been chosen and many traditional tenant-right matters will be caught; for example, growing crops, cultivations, liming, fertilising *etc.* and many other matters set out as tenant-right matters in Sched. 8 to the 1986 Act.

Where there are tenant-right matters which do not fall within the definition of routine improvements, the parties will have to make specific provision in the tenancy agreement if the tenant is to obtain compensation on quitting. The only difference between routine improvements and other physical improvements is the tenant's right to apply for arbitrator's approval retrospectively after the improvement has begun (see subs. (2)) and to enable the tenant to begin to provide the improvement during the arbitration process (see subs. (9)) which he may have to do by virtue of the nature of the improvement concerned which may have to be effected at a certain time of the year.

At first sight, requiring the landlord's written consent to tenant-right matters seems unworkable. However, it will be relevant in most cases only in the final year of the tenancy and such consents will often be in the tenancy agreement itself, either expressly or impliedly as a result of the user covenants or the husbandry obligations placed on the tenant (see the General Note to s.17(2) above).

Amount of compensation

Amount of compensation for tenant's improvement not consisting of planning permission

20.—(1) The amount of compensation payable to the tenant under section 16 of this Act in respect of any tenant's improvement shall be an amount equal to the increase attributable to the improvement in the value of the holding at the termination of the tenancy as land comprised in a tenancy.

(2) Where the landlord and the tenant have entered into an agreement in writing whereby any benefit is given or allowed to the tenant in consideration of the provision of a tenant's improvement, the amount of compensation otherwise payable in respect of that improvement shall be reduced by the proportion which the value of the benefit bears to the amount of the total cost of providing the improvement.

(3) Where a grant has been or will be made to the tenant out of public money in respect of a tenant's improvement, the amount of compensation otherwise payable in respect of that improvement shall be reduced by the proportion which the amount of the grant bears to the amount of the total cost of providing the improvement.

(4) Where a physical improvement which has been completed or a change of use which has been effected is authorised by any planning permission granted on an application made by the tenant, section 18 of this Act does not prevent any value attributable to the fact that the physical improvement or change of use is so authorised from being taken into account under this section in determining the amount of compensation payable in respect of the physical improvement or in respect of any intangible advantage obtained as a result of the change of use.

(5) This section does not apply where the tenant's improvement consists of planning permission.

DEFINITIONS
"holding": s.38(1).
"landlord": s.38(1), (5).
"planning permission": s.27 and the Town and Country Planning Act 1990, s.336(1).
"tenancy": s.38(1).
"tenant": s.38(1), (5).
"tenant's improvement": s.15.
"termination of the tenancy": s.38(1).

GENERAL NOTE

Subs. (1)
The idea of the compensation provisions is not to compensate the tenant for the cost of the improvement but to be restitutionary in nature and they are designed to ensure that the landlord is not unjustly enriched at the tenant's expense. This section works on the assumption that in many cases the benefit that the landlord would obtain from the improvement (if any) is by way of an increased rent from the holding. The landlord would obviously receive that increased rent for the length of the life of the improvement and it is that benefit or assumed benefit which must be passed to the tenant. The amount of compensation which the tenant receives, therefore, is in most cases the capitalised increase in the rental value of the holding as a result of the improvement for the length of the life of the improvement. Capitalised rental values as a basis for valuation will not be needed in the case of tenant-right matters where a clear value (for example, of the standing crop) can be seen.

The valuation takes place at the termination of the tenancy and, therefore, the tenant takes the risk of the improvement becoming unfashionable, in breach of regulations or overtaken by new technology. This also reflects the restitutionary nature of the provisions.

Subs. (2)
Reference should be made to the General Note to s.15(1) above. Unlike the position under the 1986 Act, a tenant may be compensated for improvements where he has not paid for the improvement but where it has been provided by the tenant's own efforts. This may be wide enough to compensate a tenant where the costs of materials have been met by the landlord but the labour has been put in by the tenant himself. If so, care will have to be taken by landlords to ensure that the arrangement is reduced into writing to enable them to take advantage of this subsection.

Subs. (3)
Note that for the purposes of rent review (see s.13(3) above) the grant-aided elements of the tenant's improvements still fall to be disregarded.

Subs. (4)
Unimplemented planning permissions where the physical improvement has not been completed or the change of use effected are compensated as separate items (see s.18 above). This section recognises, however, that the increase in the value of the holding attributable to an improvement may be greater where that improvement has been authorised by a valid planning consent and makes it clear that such additional value is not factored out by s.18.

Amount of compensation for planning permission

21.—(1) The amount of compensation payable to the tenant under section 16 of this Act in respect of a tenant's improvement which consists of planning permission shall be an amount equal to the increase attributable to the fact that the relevant development is authorised by the planning permission in the value of the holding at the termination of the tenancy as land comprised in a tenancy.

(2) In subsection (1) above, "the relevant development" means the physical improvement or change of use specified in the landlord's consent under section 18 of this Act in accordance with subsection (1)(b) of that section.

(3) Where the landlord and the tenant have entered into an agreement in writing whereby any benefit is given or allowed to the tenant in consideration of the obtaining of planning permission by the tenant, the amount of compensation otherwise payable in respect of that permission shall be reduced by the proportion which the value of the benefit bears to the amount of the total cost of obtaining the permission.

DEFINITIONS
"landlord": s.38(1), (5).
"planning permission": s.27 and the Town and Country Planning Act 1990, s.336(1).
"relevant development, the": subs. (2).
"tenancy": s.38(1).
"tenant": s.38(1), (5).
"tenant's improvement": s.15.
"termination of the tenancy": s.38(1).

GENERAL NOTE

Subs. (1)
Subsection (1) provides a similar method of calculating compensation to that for physical improvements and for intangible advantages other than planning permissions.

Subs. (2)
The limit in subs. (2) must be noted. The planning permission obtained may be wide enough to permit development other than the specific physical improvement or change of use authorised by the landlord's consent, but it is by reference to the consented improvement or change of use and not the planning permission itself that the increase in value is determined. Also, it is not the increase in value of the freehold as a result of the planning permission which the tenant obtains, but the increase in value of the land as land comprised in a tenancy. Landlords should ensure that their consent is narrowly drafted to the specific physical improvement or change of use required by the tenant.

Subs. (3)
See the General Note to s.20(2) above.

Settlement of claims for compensation

22.—(1) Any claim by the tenant under a farm business tenancy for compensation under section 16 of this Act shall, subject to the provisions of this section, be determined by arbitration under this section.

(2) No such claim for compensation shall be enforceable unless before the end of the period of two months beginning with the date of the termination of the tenancy the tenant has given notice in writing to his landlord of his intention to make the claim and of the nature of the claim.

(3) Where—
(a) the landlord and the tenant have not settled the claim by agreement in writing, and
(b) no arbitrator has been appointed under an agreement made since the notice under subsection (2) above was given,

either party may, after the end of the period of four months beginning with the date of the termination of the tenancy, apply to the President of the RICS for the appointment of an arbitrator by him.

(4) Where—
(a) an application under subsection (3) above relates wholly or partly to compensation in respect of a routine improvement (within the meaning of section 19 of this Act) which the tenant has provided or has begun to provide, and
(b) that application is made at the same time as an application under section 19(4) of this Act relating to the provision of that improvement,

the President of the RICS shall appoint the same arbitrator on both applications and, if both applications are made by the same person, only one fee shall be payable by virtue of section 30(2) of this Act in respect of them.

(5) Where a tenant lawfully remains in occupation of part of the holding after the termination of a farm business tenancy, references in subsections (2) and (3) above to the termination of the tenancy shall, in the case of a claim relating to that part of the holding, be construed as references to the termination of the occupation.

DEFINITIONS
"farm business tenancy": s.1.
"given": s.36.
"holding": s.38(1).
"landlord": s.38(1), (5).
"RICS, the": s.38(1).
"tenant": s.38(1), (5).
"termination of the tenancy": subs. (4), s.38(1).

GENERAL NOTE
Section 28 of the Act does not apply to arbitrations under this section (see s.28(5)), although s.30 does, and s.29 (alternative dispute resolution) is not available as s.29 only applies to disapply s.28 in circumstances where that section would otherwise apply. However, the arbitration under this section is under the Arbitration Acts 1950 to 1979.

This section provides for mandatory reference to arbitration as to the amount of compensation in the absence of agreement between the parties. Subsection (1) sets up a discrete arbitration timetable for the determination of compensation claims and is self-explanatory. This section applies to all physical improvements including routine improvements (although note subs. (4)) and to all intangible advantages including planning permissions.

Subss. (2) and (3)
The service of a notice to make a claim and the appointment of an arbitrator are familiar from the 1986 Act. There is no prescribed form for the notice although it must set out the information specified in subs. (2). Both relevant dates, for the service of the notice and for the appointment of an arbitrator in default of agreement, run from the date of the termination of the tenancy (see subs. (5) below).

Subs. (4)
This cuts down on the bureaucracy and costs where a tenant is applying at the same time for consent to routine improvements already made or started and for compensation in respect of improvements generally. In practical terms, an arbitrator will take first the question of consent for routine improvements and will then go on to deal with all questions of compensation.

Subs. (5)
This extends the period for claiming compensation in the event of the tenant holding over, for example, in the buildings until sale of the grain or on the land to take a late harvested crop.

Supplementary provisions with respect to compensation

Successive tenancies

23.—(1) Where the tenant under a farm business tenancy has remained in the holding during two or more such tenancies, he shall not be deprived of his right to compensation under section 16 of this Act by reason only that any tenant's improvement was provided during a tenancy other than the one at the termination of which he quits the holding.

(2) The landlord and tenant under a farm business tenancy may agree that the tenant is to be entitled to compensation under section 16 of this Act on the termination of the tenancy even though at that termination the tenant remains in the holding under a new tenancy.

(3) Where the landlord and the tenant have agreed as mentioned in subsection (2) above in relation to any tenancy ("the earlier tenancy"), the tenant shall not be entitled to compensation at the end of any subsequent tenancy in respect of any tenant's improvement provided during the earlier tenancy in relation to the land comprised in the earlier tenancy.

DEFINITIONS
"earlier tenancy, the": subs. (3).
"farm business tenancy": s.1.
"holding": s.38(1).
"landlord": s.38(1), (5).
"tenancy": s.38(1).
"tenant": s.38(1), (5).
"tenant's improvement": s.15.
"termination": s.38(1).
"termination of the tenancy": s.38(1).

GENERAL NOTE

Subs. (1)
Unless the tenant quits the holding on the termination of the tenancy there is no entitlement to compensation at that time (see s.16).
This provision allows the tenant to "roll over" his claim for compensation for improvements where he remains on the holding under a new tenancy. It is an important provision if, in the early days of this Act, fixed-term tenancies granted are initially going to be relatively short but may be renewed. However, as the roll over only applies between farm business tenancies, a tenant should be careful to ensure that the later tenancies are farm business tenancies and have not, for example, slipped into the Landlord and Tenant Act 1954, Pt. II, by reason of the fact that they are diversified estates in respect of which it can no longer be said that they are primarily or wholly agricultural at the outset (see the General Note to s.1 above).
"Holding" (see s.38(1)) means the aggregate of land comprised in the tenancy, and this subsection will only, therefore, apply where the land in the two or more tenancies is the same, although a *de minimis* rule may apply. It is the land which must be the same and not the terms of the subsequent tenancies. This subsection preserves the tenant's claim not only where there has been a technical surrender and regrant but also where he continues to occupy under the terms of a new tenancy and where he cannot obtain the landlord's agreement or does not wish to obtain the landlord's agreement under subs. (2) below.

Subss. (2) and (3)
By virtue of s.16 above, a tenant's right to compensation only arises on the termination of the tenancy upon the tenant quitting the holding. If he remains, therefore, the entitlement does not

arise. The route by which the tenant can take his compensation at the end of the earlier tenancy is by obtaining the agreement of the landlord in accordance with this subsection. If the landlord refuses to agree, the tenant's only option is to refuse the new tenancy and quit the holding to obtain compensation. There is no fallback of arbitration for the refusal of agreement under subs. (2). Where there is a risk that the increase in value of the holding as a result of the improvement will be significantly less at the end of a later tenancy when the tenant actually quits, whether as a result of the age of the improvement, technological advances or changes in a relevant regulatory framework, a tenant should seek the agreement of the landlord to take compensation early. In circumstances where it is likely, however, to be in the interests of a tenant to want to take his compensation early, it is unlikely to be in the interests of the landlord. However, this issue will not be treated in isolation but as part and parcel of the negotiations for the new tenancy.

Subs. (3)
Subsection (3) makes it clear that the tenant is not entitled to be compensated twice for the same improvement.

Resumption of possession of part of holding

24.—(1) Where—
(a) the landlord under a farm business tenancy resumes possession of part of the holding in pursuance of any provision of the tenancy, or
(b) a person entitled to a severed part of the reversionary estate in a holding held under a farm business tenancy resumes possession of part of the holding by virtue of a notice to quit that part given to the tenant by virtue of section 140 of the Law of Property Act 1925,

the provisions of this Part of this Act shall, subject to subsections (2) and (3) below, apply to that part of the holding (in this section referred to as "the relevant part") as if it were a separate holding which the tenant had quitted in consequence of a notice to quit and, in a case falling within paragraph (b) above, as if the person resuming possession were the landlord of that separate holding.

(2) The amount of compensation payable to the tenant under section 16 of this Act in respect of any tenant's improvement provided for the relevant part by the tenant and not consisting of planning permission shall, subject to section 20(2) to (4) of this Act, be an amount equal to the increase attributable to the tenant's improvement in the value of the original holding on the termination date as land comprised in a tenancy.

(3) The amount of compensation payable to the tenant under section 16 of this Act in respect of any tenant's improvement which consists of planning permission relating to the relevant part shall, subject to section 21(3) of this Act, be an amount equal to the increase attributable to the fact that the relevant development is authorised by the planning permission in the value of the original holding on the termination date as land comprised in a tenancy.

(4) In a case falling within paragraph (a) or (b) of subsection (1) above, sections 20 and 21 of this Act shall apply on the termination of the tenancy, in relation to the land then comprised in the tenancy, as if the reference in subsection (1) of each of those sections to the holding were a reference to the original holding.

(5) In subsections (2) to (4) above—
"the original holding" means the land comprised in the farm business tenancy—
(a) on the date when the landlord gave his consent under section 17 or 18 of this Act in relation to the tenant's improvement, or
(b) where approval in relation to the tenant's improvement was given by an arbitrator, on the date on which that approval was given,
"the relevant development", in relation to any tenant's improvement which consists of planning permission, has the meaning given by section 21(2) of this Act, and

"the termination date" means the date on which possession of the relevant part was resumed.

DEFINITIONS
"farm business tenancy": s.1.
"holding": s.38(1).
"landlord": s.38(1), (5).
"original holding, the": subs. (5).
"planning permission": s.27 and the Town and Country Planning Act 1990, s.336(1).
"relevant development, the": subs. (5), s.21(2).
"relevant part, the": subs. (1).
"tenancy": s.38(1).
"tenant": s.38(1), (5).
"tenant's improvement": s.15.
"termination date, the": subs. (5).
"termination of the tenancy": s.38(1).

GENERAL NOTE
Section 74 of the 1986 Act contains similar provisions as found in this section but does not require the assessment of compensation by reference to the original holding (see below).

Subs. (1)
This subsection only applies where a landlord has instigated the resumption of part of the holding either as a result of the exercise of a break clause or by service of a notice to quit under s.140(2) of the Law of Property Act 1925 by reason of being the owner of a severed reversionary estate. On s.140(2) see the General Note to s.5 above.
This subsection ensures that the tenant's entitlement to compensation arises when the landlord resumes possession of that part.

Subs. (2)
The entitlement to compensation is in respect of improvements "provided for" the part of the holding being taken back. This may cover improvements on adjoining land which serve only the part taken back, for example, storage facilities. The compensation is assessed not by reference to the increase in value of the part taken back, which could give a distorted picture, but by reference to the land which was within the tenancy at the date that the consent or arbitrator's approval was given for the improvement.

Subs. (3)
This provides a similar provision for planning permission save that the planning permission must relate to the part taken back.

Subs. (4)
This ensures that on the termination of the tenancy (which will then only include the balance of the land) compensation is assessed by reference to the land which was in the tenancy at the date that the consent or arbitrator's approval was given for the improvement.

Compensation where reversionary estate in holding is severed

25.—(1) Where the reversionary estate in the holding comprised in a farm business tenancy is for the time being vested in more than one person in several parts, the tenant shall be entitled, on quitting the entire holding, to require that any compensation payable to him under section 16 of this Act shall be determined as if the reversionary estate were not so severed.

(2) Where subsection (1) applies, the arbitrator shall, where necessary, apportion the amount awarded between the persons who for the purposes of this Part of this Act together constitute the landlord of the holding, and any additional costs of the award caused by the apportionment shall be directed by the arbitrator to be paid by those persons in such proportions as he shall determine.

DEFINITIONS
"farm business tenancy": s.1.
"holding": s.38(1).

"landlord": s.38(1), (5).
"tenant": s.38(1), (5).

GENERAL NOTE
This provision replicates s.75 of the 1986 Act.

Subs. (1)
This is an enabling provision. The tenant can, if it suits him to do so, have his compensation claim separately assessed for each severed part of the reversion as a tenant can under s.75 of the 1986 Act. However, if the assessment is under the provisions of s.24, above, the improvement will nevertheless be assessed by reference to the holding as it was at the time of the consent or the arbitrator's approval to the improvement. Section 24 will not apply in all cases as it only operates if the landlord exercises a break clause or if the owner of a severed part of a reversionary estate serves a notice to quit under s.140(2) of the Law of Property Act 1925. Where the tenant serves a notice to quit on an estate with a severed reversion, he can choose to treat each severed part as a separate holding and have the improvement assessed by reference to the increase in value of that part (and not the whole) if it is advantageous to him to do so.

Subs. (2)
There is no guidance for arbitrators on how such an apportionment should be made.

Extent to which compensation recoverable under agreements

26.—(1) In any case for which apart from this section the provisions of this Part of this Act provide for compensation, a tenant shall be entitled to compensation in accordance with those provisions and not otherwise, and shall be so entitled notwithstanding any agreement to the contrary.

(2) Nothing in the provisions of this Part of this Act, apart from this section, shall be construed as disentitling a tenant to compensation in any case for which those provisions do not provide for compensation.

DEFINITIONS
"tenant": s.38(1), (5).

GENERAL NOTE
This section prevents contracting out of Pt. III of the Act if the Act provides for compensation for the particular item or matter. The parties are unable to provide in the tenancy agreement for no compensation and cannot provide for a different basis of assessment or, for example, that the improvement be written down over a number of years. There is nothing to prevent the parties from agreeing compensation for items or matters not covered by the Act, for example, in connection with items of tenant-right which do not fall within the definition of routine improvements.

Interpretation of Part III

27. In this Part of this Act, unless the context otherwise requires—
"planning permission" has the meaning given by section 336(1) of the Town and Country Planning Act 1990;
"tenant's improvement", and references to the provision of such an improvement, have the meaning given by section 15 of this Act.

PART IV

MISCELLANEOUS AND SUPPLEMENTAL

Resolution of disputes

Resolution of disputes

28.—(1) Subject to subsections (4) and (5) below and to section 29 of this Act, any dispute between the landlord and the tenant under a farm business tenancy, being a dispute concerning their rights and obligations under this Act, under the terms of the tenancy or under any custom, shall be determined by arbitration.

(2) Where such a dispute has arisen, the landlord or the tenant may give notice in writing to the other specifying the dispute and stating that, unless before the end of the period of two months beginning with the day on which the notice is given the parties have appointed an arbitrator by agreement, he proposes to apply to the President of the RICS for the appointment of an arbitrator by him.

(3) Where a notice has been given under subsection (2) above, but no arbitrator has been appointed by agreement, either party may, after the end of the period of two months referred to in that subsection, apply to the President of the RICS for the appointment of an arbitrator by him.

(4) Subsection (1) above does not affect the jurisdiction of the courts, except to the extent provided by section 4(1) of the Arbitration Act 1950 (staying of court proceedings where there is submission to arbitration), as applied to statutory arbitrations by section 31 of that Act.

(5) Subsections (1) to (3) above do not apply in relation to—
  (a) the determination of rent in pursuance of a statutory review notice (as defined in section 10(1) of this Act),
  (b) any case falling within section 19(1) of this Act, or
  (c) any claim for compensation under Part III of this Act.

DEFINITIONS
  "farm business tenancy": s.1.
  "give/given": s.36.
  "landlord": s.38(1), (5).
  "RICS, the": s.38(1).
  "statutory review notice": s.10(1).
  "tenancy": s.38(1).
  "tenant": s.38(1), (5).

GENERAL NOTE

Subs. (1)
Section 28 applies to arbitrations for disputes other than in connection with rent review, consent to improvements or compensation for improvements each of which have their own procedure set out in the Act (see subs. (5) below). Section 28 arbitrations are a fallback for parties who have not agreed an alternative dispute resolution mechanism or, if they have agreed one, have not agreed on the particular occasion of the dispute in question that it will be used (see s.29 below). Arbitrations under this section will be under the Arbitration Acts 1950 to 1979 and not under any discrete code. This is a departure from the position under the 1986 Act where all arbitrations are carried out under the provisions of a discrete statutory code set out in the Act itself (see s.84 and Sched. 11). Schedule 11 to the 1986 Act provides a highly regulated code, specific procedures and mandatory time-limits, and the change over to the Arbitration Acts procedure, which leaves far more in the hands of the parties and the arbitrator, will be a significant change for those involved in agricultural arbitrations to adapt to.

Apart from the provisions of the Arbitration Acts themselves and case law decided under those Acts, the only regulation of arbitrations under s.28 or under the specific arbitrations relating to rent reviews, consent to improvements and compensation for improvements is contained in s.30.

Discussion of the Arbitration Acts procedures is beyond the scope of this annotation and reference should be made to the standard books on commercial arbitrations (for example, *Handbook of Arbitration Practice*, Bernstein & Wood (1993, Sweet & Maxwell); *Commercial Arbitration*, Mustill & Boyd (2nd edition, Butterworths)). However, one or two points showing the difference between 1986 Act arbitrations and Arbitration Acts arbitrations are worth mentioning.

(1) Under the 1986 Act the parties are required to deliver a statement of their case, with all of the necessary particulars, to the arbitrator within 35 days from the date of his appointment (Sched. 11, para. 7). The time-limit is mandatory and inflexible. Under the Arbitration Acts, how the case is pleaded out is entirely in the hands of the arbitrator, not only in terms of whether he requires pleadings or statements of case but also in terms of whether those documents are to be exchanged or produced sequentially and in terms of the timetable for the pleadings. Furthermore, failure to comply with the timetable set by an arbitrator under the Arbitration Acts can only effectively be dealt with by the arbitrator if an application is made by one of the parties

under s.5 of the Arbitration Act 1979 (c. 42) to the High Court to allow the arbitrator to extend his powers to those of a High Court Judge. Whilst the arbitrator may himself apply under s.5, it is rare and it is more likely that the party not in default will make the application.

(2) The method by which the arbitrator deals with points of law is different. Under the 1986 Act, the case stated procedure enabled an arbitrator, at any stage of the proceedings, to state a case for the opinion of the county court on any question of law or any question relating to the arbitrator's jurisdiction (see s.84(1) of and Sched. 11, para. 26 to the 1986 Act). Under the Arbitration Acts, however, the case stated procedure does not exist. In the MAFF & WOAD consultation document of September 1992, which set out detailed proposals for this legislation, the suggestion was that, whilst the discrete arbitration code of the 1986 Act would be disposed of, the ability of the arbitrator to refer points of law to the courts would be preserved.

Under the Arbitration Acts, the High Court may determine any question of law arising in the course of the reference on the application of any of the parties with the consent of the arbitrator or on the application of any of the parties with the consent of all of the other parties (see s.2 of the Arbitration Act 1979). The ability of the arbitrator to consent without the consent of all other parties is limited (see s.2(2) of the Arbitration Act 1979).

(3) The requirement under the 1986 Act that an award be delivered within 56 days of the date of the appointment of an arbitrator (Sched. 11, para. 14) is not repeated in this Act and does not appear in the Arbitration Acts. The award under the 1986 Act is in a prescribed form. Under the Arbitration Acts, there are no legal requirements as to the form of the award and the absence of reasons does not invalidate the award. However, if a reasoned award is requested by both parties, the arbitrator must give reasons.

Subs. (2)

There is no prescribed form for the Notice, although the prescribed information set out in the subsection must be contained in the Notice.

Subs. (3)

The application must be in writing (see s.30(2) and (3)). In making the appointment of the arbitrator, the President of the RICS is acting in an administrative and not a judicial capacity (see *Ramsey v. McClaren* [1936] S.L.T. 35). Accordingly if either party disputes the validity of the arbitrator's appointment, the matter should be raised with the arbitrator himself and not with the President of the RICS.

Subs. (4)

This subsection makes it clear that the use of the mandatory "shall" in subs. (1) above does not operate so as to oust the jurisdiction of the courts save to the extent set out in that subsection. Section 4(1) of the Arbitration Act 1950 (c. 27) provides that a court may stay any proceedings commenced in the court in respect of any matter which falls within the scope of an arbitration agreement or (by s.31 of the Arbitration Act 1950) within the scope of a statutory arbitration. The courts will stay the proceedings if satisfied that there is no sufficient reason why the matter should not be referred to arbitration and that the applicant who is attempting to stay the court proceedings was and is ready, willing and able to do all things necessary for the proper conduct of the arbitration. If neither party applies for a stay of the proceedings commenced in the court, the hearing will go ahead and hence the parties can agree to go to court rather than to arbitration.

Subs. (5)

This sets out the three specific references to arbitration for rent review, approval of the arbitrator to an improvement where a landlord has refused or failed to give consent, and a claim for compensation for tenant's improvements on quitting the holding.

Those specific arbitrations will still be under the Arbitration Acts 1950 to 1979 but have their own timetables set out in the relevant sections of this Act.

Cases where right to refer claim to arbitration under section 28 does not apply

29.—(1) Section 28 of this Act does not apply in relation to any dispute if—
    (a) the tenancy is created by an instrument which includes provision for disputes to be resolved by any person other than—

(i) the landlord or the tenant, or
(ii) a third party appointed by either of them without the consent or concurrence of the other, and
(b) either of the following has occurred—
(i) the landlord and the tenant have jointly referred the dispute to the third party under the provision, or
(ii) the landlord or the tenant has referred the dispute to the third party under the provision and notified the other in writing of the making of the reference, the period of four weeks beginning with the date on which the other was so notified has expired and the other has not given a notice under section 28(2) of this Act in relation to the dispute before the end of that period.

(2) For the purposes of subsection (1) above, a term of the tenancy does not provide for disputes to be "resolved" by any person unless that person (whether or not acting as arbitrator) is enabled under the terms of the tenancy to give a decision which is binding in law on both parties.

DEFINITIONS
"given": s.36.
"landlord": s.38(1), (5).
"tenancy": s.38(1).
"tenant": s.38(1), (5).

GENERAL NOTE

Subs. (1)
The arbitration provisions set out in s.28 above are designed as a fallback in the absence of some other agreed mechanism for dispute resolution.
This section allows the parties to choose an alternative dispute resolution mechanism but provides either party with the ability to opt back into arbitration for any particular dispute despite an agreement for an alternative procedure. The alternative dispute resolution procedure only applies to oust s.28 of the Act and, therefore, does not apply to oust arbitration in connection with the rent review arbitration, the consent to improvement arbitration or the compensation for tenant's improvement arbitration referred to in s.28(5). However, so far as rent review is concerned, the ability of the parties to follow through their own rent review formula not within s.9, by reference to an expert, has already been discussed (see s.12(b) above).
The ability to avoid arbitration only arises where the alternative dispute mechanism is contained in the tenancy agreement itself and where that tenancy agreement is in writing. Unlike the position in connection with the rent review procedure in s.12, this section does not prohibit the parties from agreeing to a contractual arbitration. An express arbitration clause can, of course, circumscribe the jurisdiction of the arbitrator and it may be advantageous in particular circumstances for the parties to do that.
Even where the parties choose to insert into their tenancy agreement an alternative dispute resolution mechanism, the parties must agree in connection with each and every dispute that that mechanism will be utilised. It is open to either party in connection with any dispute, to opt out and to force the matter back into the statutory arbitration framework. The method by which a party does that is set out in subs. (1)(b)(ii): he gives a notice within the timetable set out in that subsection that he wishes the matter to be referred to arbitration.

Subs. (2)
As this "opt out" is only available where the person to be appointed is able to give a decision which is binding in law on both parties, the position is that where the parties have jointly appointed a third party to determine the dispute or where the timetable as set out in subs. (1)(b)(ii) has expired, the award made by that third party will be binding on the parties. Whether or not the parties can abandon the alternative dispute mechanism at a stage before the award is given depends upon the terms of the contractual reference itself. However, the parties will then be left in some difficulty in the determination of that dispute as they cannot opt back into s.28 once their agreed third party has been appointed. Any form of alternative dispute resolution

mechanism or mediation which does not specifically have the power to bind the parties can be utilised as the arbitration provisions are only a fallback in the absence of agreement. However, such facilities do not have the power to bind the parties and do not fall within s.29 so as to prevent the parties from ignoring any decision and then referring the matter to arbitration.

General provisions applying to arbitrations under Act

30.—(1) Any matter which is required to be determined by arbitration under this Act shall be determined by the arbitration of a sole arbitrator.

(2) Any application under this Act to the President of the RICS for the appointment of an arbitrator by him must be made in writing and must be accompanied by such reasonable fee as the President may determine in respect of the costs of making the appointment.

(3) Where an arbitrator appointed for the purposes of this Act dies or is incapable of acting and no new arbitrator has been appointed by agreement, either party may apply to the President of the RICS for the appointment of a new arbitrator by him.

DEFINITIONS
  "RICS, the": s.37(1).

GENERAL NOTE

Subs. (2)
  See the General Note to s.28(3) above.

Miscellaneous

Mortgages of agricultural land

31.—(1) Section 99 of the Law of Property Act 1925 (leasing powers of mortgagor and mortgagee in possession) shall be amended in accordance with subsections (2) and (3) below.

(2) At the beginning of subsection (13), there shall be inserted "Subject to subsection (13A) below,".

(3) After that subsection, there shall be inserted—
  "(13A) Subsection (13) of this section—
  (a) shall not enable the application of any provision of this section to be excluded or restricted in relation to any mortgage of agricultural land made after 1st March 1948 but before 1st September 1995, and
  (b) shall not enable the power to grant a lease of an agricultural holding to which, by virtue of section 4 of the Agricultural Tenancies Act 1995, the Agricultural Holdings Act 1986 will apply, to be excluded or restricted in relation to any mortgage of agricultural land made on or after 1st September 1995.
  (13B) In subsection (13A) of this section—
    'agricultural holding' has the same meaning as in the Agricultural Holdings Act 1986; and
    'agricultural land' has the same meaning as in the Agriculture Act 1947."

(4) Paragraph 12 of Schedule 14 to the Agricultural Holdings Act 1986 (which excludes the application of subsection (13) of section 99 of the Law of Property Act 1925 in relation to a mortgage of agricultural land and is super-

seded by the amendments made by subsections (1) to (3) above) shall cease to have effect.

DEFINITIONS
"agricultural holding": subs. (3).
"agricultural land": subs. (3).

GENERAL NOTE
Section 99 of the Law of Property Act 1925 confers upon a mortgagor in possession the power to make such leases as are authorised by the section in such a way as bind a prior mortgagee. Authorised leases include agricultural or occupation leases not exceeding 50 years (see s.99(3)(i)).

Such a lease must satisfy the requirements of s.99(5)–(7) namely: it must take effect in possession within 12 months; it must reserve the best rent reasonably obtainable, regard being had to the circumstances of the case but without any fine being taken; and it must contain a covenant for the payment of rent and a condition for re-entry if the rent is unpaid for a period specified not exceeding 30 days.

The section authorises agreements for leases and oral lettings and, in such cases, the absence of a condition for re-entry and the failure to comply with the requirement for the delivery of a counterpart lease to the mortgagee (see s.99(8)) will not affect the validity of the lease or its ability to bind the mortgagee (*Rhodes v. Dalby* [1971] 1 W.L.R. 1325).

Section 99(13) provides that the s.99 powers of leasing shall not apply if there is a clause in the mortgage document excluding those powers. All standard mortgage documentation excludes the powers conferred on the mortgagor by s.99. However, for mortgages of agricultural land made after March 1, 1948, Sched. 14, para. 12 to the Agricultural Holdings Act 1986 provides that the powers given to the mortgagor cannot be excluded.

As a tenancy granted under the 1986 Act could devalue the land by as much as 50 per cent, the impact of such a grant on the mortgagee's security, where the mortgagee has lent against the freehold vacant possession value of the land, is acute. The grant of a long fixed-term farm business tenancy would also have a devaluing effect on the land and, it should be noted, that tenancies with residential security of tenure or under the business tenancy regime in the Landlord and Tenant Act 1954, Pt. II, could also be granted. It is the nature of the mortgage and not the tenancy which brings Sched. 14, para. 12 to the 1986 Act into play.

In circumstances where a tenancy is granted where the mortgagor is already in financial difficulties, whilst the tenancy may bind the mortgagee, it may be set aside by an application under s.423 of the Insolvency Act 1986 (c. 45) as a transaction at an undervalue even where the best rent is being paid (see the Court of Appeal decision in *Agricultural Mortgage Corp. v. Woodward* [1995] 04 E.G. 155).

Section 31 introduces an amendment to s.99 of the Law of Property Act 1925 which replaces Sched. 14, para 12, to the 1986 Act which, on September 1, 1995, shall cease to have effect. The new s.99(13A) which is introduced to the 1925 Act means that, in relation to any mortgage of agricultural land made between March 1, 1948 and September 1, 1995, the powers in s.99 still cannot be excluded. This fits with the government's stated policy that the Act should not be retrospective in any respect. Any mortgage made on or after September 1, 1995 can validly contain a prohibition on leasing, regardless of whether or not it is a mortgage of agricultural land.

The cut off date is in relation to the date of the mortgage and not the date of the tenancy agreement. If a mortgage of agricultural land is already in existence at the time of the coming into force of the Act, the mortgagee still cannot contract out of the s.99 powers of leasing. What may happen is that lending institutions will review their agricultural lending portfolios and insist on new security being taken on or after September 1, 1995. Whether or not these moves will be attacked by the courts as being against public policy remains to be seen and probably depends as much on the circumstances surrounding the grant of the tenancy as the motives of the mortgagee.

Power of limited owners to give consents etc.

32. The landlord under a farm business tenancy, whatever his estate or interest in the holding, may, for the purposes of this Act, give any consent, make any agreement or do or have done to him any other act which he might give, make, do or have done to him if he were owner in fee simple or, if his interest is an interest in a leasehold, were absolutely entitled to that leasehold.

DEFINITIONS
"farm business tenancy": s.1.
"holding": s.38(1).
"landlord": s.38(1), (5).

GENERAL NOTE
This replicates s.88 of the 1986 Act.

Power to apply and raise capital money

33.—(1) The purposes authorised by section 73 of the Settled Land Act 1925 (either as originally enacted or as applied in relation to trusts for sale by section 28 of the Law of Property Act 1925) or section 26 of the Universities and College Estates Act 1925 for the application of capital money shall include—
   (a) the payment of expenses incurred by a landlord under a farm business tenancy in, or in connection with, the making of any physical improvement on the holding,
   (b) the payment of compensation under section 16 of this Act, and
   (c) the payment of the costs, charges and expenses incurred by him on a reference to arbitration under section 19 or 22 of this Act.

(2) The purposes authorised by section 71 of the Settled Land Act 1925 (either as originally enacted or as applied in relation to trusts for sale by section 28 of the Law of Property Act 1925) as purposes for which money may be raised by mortgage shall include the payment of compensation under section 16 of this Act.

(3) Where the landlord under a farm business tenancy—
   (a) is a tenant for life or in a fiduciary position, and
   (b) is liable to pay compensation under section 16 of this Act,
he may require the sum payable as compensation and any costs, charges and expenses incurred by him in connection with the tenant's claim under that section to be paid out of any capital money held on the same trusts as the settled land.

(4) In subsection (3) above—
   "capital money" includes any personal estate held on the same trusts as the land; and
   "settled land" includes land held on trust for sale or vested in a personal representative.

DEFINITIONS
"capital money": subs. (4).
"farm business tenancy": s.1.
"holding": s.38(1).
"landlord": s.38(1), (5).
"settled land": subs. (4).
"tenant": s.38(1), (5).

GENERAL NOTE
Section 89 of the 1986 Act contains similar provisions. Certain problems with the provisions in s.89 have been avoided in s.33. The provisions of the Settled Land Act 1925 (c. 18) on the permitted uses of capital money and on the repayment of capital money spent on certain improvements by instalments out of income, are designed to maintain the correct balance between the interest of those persons entitled for life (and hence to income) and those persons entitled in remainder (and hence to capital). Section 89 of the 1986 Act specifically provided for improvements listed in Sched. 7 to the 1986 Act, when paid for out of capital monies, not to have to be replaced out of income. As Sched. 7 includes repairs to fixed equipment, the remaindermen essentially end up paying for repairs which does not accord with the spirit of the Settled Land Act 1925 provisions.

A further problem with s.89 is that a tenant for life or a trustee for sale who paid out compensation for improvements himself had no right of recompense directly from the settlement or trust funds (see *Duke of Wellington's Parliamentary Estates, Re*; *King v. Wellesley* [1972] Ch. 374).

Subs. (1)
This subsection extends s.73 of the Settled Land Act 1925 to enable capital monies to be spent on the matters set out in subs. (1)(a) to (c) but makes no mention of repayment out of income. Reference is therefore now made direct to the provisions of the Settled Land Act 1925 itself (ss.83 to 89 and Sched. 3) which set out those improvements in respect of which instalment payments out of income do not need to be made, where they must and where the trustees or the court have a discretion. However, s.84 and Sched. 3 deal with the provision of money "... in or towards payment for any improvement" and it is arguable whether that would cover a payment of compensation to the tenant for the provision of such an improvement. It is submitted that it does.

Subs. (2)
This allows the monies to be paid in compensation to a tenant to be raised on mortgage.

Subs. (3)
This deals with the problem identified in the *Duke of Wellington's Parliamentary Estates* (see above) and enables the tenant for life or the trustee to obtain repayment direct from the relevant fund.

Estimation of best rent for purposes of Acts and other instruments

34.—(1) In estimating the best rent or reservation in the nature of rent of land comprised in a farm business tenancy for the purposes of a relevant instrument, it shall not be necessary to take into account against the tenant any increase in the value of that land arising from any tenant's improvements.

(2) In subsection (1) above—
"a relevant instrument" means any Act of Parliament, deed or other instrument which authorises a lease to be made on the condition that the best rent or reservation in the nature of rent is reserved;
"tenant's improvement" has the meaning given by section 15 of this Act.

DEFINITIONS
"farm business tenancy": s.1.
"relevant instrument, a": subs. (2).
"tenant": s.38(1), (5).
"tenant's improvement": subs. (2), s.15.

GENERAL NOTE
There are several provisions which require best rent to be taken in respect of a lease. See, for example, the following provisions.
(1) Section 42 of the Settled Land Act 1925 confers powers on a tenant for life to grant leases provided that such leases reserve the best rent. By s.28 of the Law of Property Act 1925, this power is also conferred on trustees for sale.
(2) Section 54 of the Law of Property Act 1925 provides that leases which do not exceed three years can be oral or in writing provided that they are at the best rent. Otherwise, such leases would have to be executed by deed in accordance with s.52 of the Law of Property Act 1925.
(3) Section 99 of the Law of Property Act 1925 confers upon a mortgagor in possession powers of leasing so as to bind a mortgagee where the lease reserves the best rent (see the General Note to s.31 above).
As all of the examples set out above require the lease simply to reserve the best rent or be at the best rent, this section can only be relevant where a new tenancy has been granted to a tenant who is already in place as the question of best rent is only relevant at the commencement of the tenancy.

Preparation of documents etc. by valuers and surveyors

35.—(1) Section 22 of the Solicitors Act 1974 (unqualified person not to prepare certain instruments) shall be amended as follows.

(2) In subsection (2), after paragraph (ab) there shall be inserted—
"(ac) any accredited person drawing or preparing any instrument—
(i) which creates, or which he believes on reasonable grounds will create, a farm business tenancy (within the meaning of the Agricultural Tenancies Act 1995), or
(ii) which relates to an existing tenancy which is, or which he believes on reasonable grounds to be, such a tenancy;".
(3) In subsection (3A), immediately before the definition of "registered trade mark agent" there shall be inserted—
"'accredited person' means any person who is—
(a) a Full Member of the Central Association of Agricultural Valuers,
(b) an Associate or Fellow of the Incorporated Society of Valuers and Auctioneers, or
(c) an Associate or Fellow of the Royal Institution of Chartered Surveyors;".

GENERAL NOTE

The likely increase in fixed-term tenancies as a result of this legislation will lead to an increase in the number of agricultural tenancies which will need to be executed by deed. Section 52 of the Law of Property Act 1925 requires all conveyances of land or of any interest in land to be by deed. Failure to comply with s.52 renders the conveyance void for the purposes of conveying or creating a legal estate. As a result of exceptions in s.54 of the Law of Property Act 1925, if the lease is to take effect in possession for a term not exceeding three years and at the best rent, it can be in writing or oral and need not be executed by deed.

Prior to the coming into force of this Act it was common for land agents and surveyors to prepare tenancy agreements for the grant of tenancies protected by the 1986 Act, giving effectively lifetime security, and this section recognises that expertise.

Under s.22 of the Solicitors Act 1974 it is a criminal offence for persons who are not solicitors to prepare deeds relating to real property if done for a fee, or for gain or reward. Section 22(2) sets out a list of exceptions for certain classes of accredited persons to enable them to prepare certain classes of documents. This amendment extends that list to enable those persons listed in the amendment set out in subs. (3) to prepare deeds relating to farm business tenancies, whether creating such a tenancy or dealing with surrenders or assignments.

Supplemental

Service of notices

36.—(1) This section applies to any notice or other document required or authorised to be given under this Act.

(2) A notice or other document to which this section applies is duly given to a person if—
 (a) it is delivered to him,
 (b) it is left at his proper address, or
 (c) it is given to him in a manner authorised by a written agreement made, at any time before the giving of the notice, between him and the person giving the notice.

(3) A notice or other document to which this section applies is not duly given to a person if its text is transmitted to him by facsimile or other electronic means otherwise than by virtue of subsection (2)(c) above.

(4) Where a notice or other document to which this section applies is to be given to a body corporate, the notice or document is duly given if it is given to the secretary or clerk of that body.

(5) Where—
 (a) a notice or other document to which this section applies is to be given to a landlord under a farm business tenancy and an agent or servant of his is responsible for the control of the management of the holding, or
 (b) such a document is to be given to a tenant under a farm business tenancy and an agent or servant of his is responsible for the carrying on of a business on the holding,

the notice or document is duly given if it is given to that agent or servant.

(6) For the purposes of this section, the proper address of any person to whom a notice or other document to which this section applies is to be given is—
  (a) in the case of the secretary or clerk of a body corporate, the registered or principal office of that body, and
  (b) in any other case, the last known address of the person in question.

(7) Unless or until the tenant under a farm business tenancy has received—
  (a) notice that the person who before that time was entitled to receive the rents and profits of the holding ("the original landlord") has ceased to be so entitled, and
  (b) notice of the name and address of the person who has become entitled to receive the rents and profits,
any notice or other document given to the original landlord by the tenant shall be deemed for the purposes of this Act to have been given to the landlord under the tenancy.

DEFINITIONS
  "farm business tenancy": s.1.
  "holding": s.38(1).
  "landlord": s.38(1), (5).
  "original landlord, the": subs. (7).
  "proper address": subs. (6).
  "tenancy": s.38(1).
  "tenant": s.38(1), (5).

GENERAL NOTE

Subs. (2)
A notice may presumably be "delivered" or "left" even if sent through in the postal system. Any agreement for service by a different method must be in writing. Such an agreement can be included in the tenancy agreement itself but, if the notices are preliminary notices to comply with the notice conditions (see s.1(4) above) the agreement will have to precede the tenancy agreement.

Subs. (3)
Electronic means of communication will mean, in addition to facsimile, e-mail or telex. Where service is to take effect between agents in accordance with subs. (5) below, an agreement to allow such methods of service may be useful although care will have to be taken to ensure that evidential safeguards are built into the agreement so that receipt of a notice by these means has to be acknowledged within a specified timescale.

Subs. (5)
If subs. (5) is to be relied upon, it will be prudent to ensure that there is agreement between the parties at the outset, updated as necessary, as to whether a particular agent falls within the description set out in paras. (a) or (b).

Subs. (7)
Note also that ss.47 and 48 of the Landlord and Tenant Act 1987 (c. 31) apply to agricultural holdings (see *Dallhold Estates (U.K.) Pty (In Administration) v. Lindsey Trading Properties* [1994] 17 E.G. 148).

Crown land

37.—(1) This Act shall apply in relation to land in which there subsists, or has at any material time subsisted, a Crown interest as it applies in relation to land in which no such interest subsists or has ever subsisted.

(2) For the purposes of this Act—
  (a) where an interest belongs to Her Majesty in right of the Crown and forms part of the Crown Estate, the Crown Estate Commissioners shall be treated as the owner of the interest,
  (b) where an interest belongs to Her Majesty in right of the Crown and does not form part of the Crown Estate, the government department

having the management of the land or, if there is no such department, such person as Her Majesty may appoint in writing under the Royal Sign Manual shall be treated as the owner of the interest,
(c) where an interest belongs to Her Majesty in right of the Duchy of Lancaster, the Chancellor of the Duchy shall be treated as the owner of the interest,
(d) where an interest belongs to a government department or is held in trust for Her Majesty for the purposes of a government department, that department shall be treated as the owner of the interest, and
(e) where an interest belongs to the Duchy of Cornwall, such person as the Duke of Cornwall or the possessor for the time being of the Duchy of Cornwall appoints shall be treated as the owner of the interest and, in the case where the interest is that of landlord, may do any act or thing which a landlord is authorised or required to do under this Act.

(3) If any question arises as to who is to be treated as the owner of a Crown interest, that question shall be referred to the Treasury, whose decision shall be final.

(4) In subsections (1) and (3) above "Crown interest" means an interest which belongs to Her Majesty in right of the Crown or of the Duchy of Lancaster or to the Duchy of Cornwall, or to a government department, or which is held in trust for Her Majesty for the purposes of a government department.

(5) Any compensation payable under section 16 of this Act by the Chancellor of the Duchy of Lancaster may be raised and paid under section 25 of the Duchy of Lancaster Act 1817 (application of monies) as an expense incurred in improvement of land belonging to Her Majesty in right of the Duchy.

(6) In the case of land belonging to the Duchy of Cornwall, the purposes authorised by section 8 of the Duchy of Cornwall Management Act 1863 (application of monies) for the advancement of parts of such gross sums as are there mentioned shall include the payment of compensation under section 16 of this Act.

(7) Nothing in subsection (6) above shall be taken as prejudicing the operation of the Duchy of Cornwall Management Act 1982.

DEFINITIONS
"Crown interest": subs. (4).
"landlord": s.38(1), (5).

Interpretation

38.—(1) In this Act, unless the context otherwise requires—
"agriculture" includes horticulture, fruit growing, seed growing, dairy farming and livestock breeding and keeping, the use of land as grazing land, meadow land, osier land, market gardens and nursery grounds, and the use of land for woodlands where that use is ancillary to the farming of land for other agricultural purposes, and "agricultural" shall be construed accordingly;
"building" includes any part of a building;
"fixed term tenancy" means any tenancy other than a periodic tenancy;
"holding", in relation to a farm business tenancy, means the aggregate of the land comprised in the tenancy;
"landlord" includes any person from time to time deriving title from the original landlord;
"livestock" includes any creature kept for the production of food, wool, skins or fur or for the purpose of its use in the farming of land;
"the RICS" means the Royal Institution of Chartered Surveyors;
"tenancy" means any tenancy other than a tenancy at will, and includes a sub-tenancy and an agreement for a tenancy or sub-tenancy;

"tenant" includes a sub-tenant and any person deriving title from the original tenant or sub-tenant;
"termination", in relation to a tenancy, means the cesser of the tenancy by reason of effluxion of time or from any other cause.

(2) References in this Act to the farming of land include references to the carrying on in relation to land of any agricultural activity.

(3) A tenancy granted pursuant to a contract shall be taken for the purposes of this Act to have been granted when the contract was entered into.

(4) For the purposes of this Act a tenancy begins on the day on which, under the terms of the tenancy, the tenant is entitled to possession under that tenancy; and references in this Act to the beginning of the tenancy are references to that day.

(5) The designations of landlord and tenant shall continue to apply until the conclusion of any proceedings taken under this Act in respect of compensation.

GENERAL NOTE

Definitions have been considered where the words or phrases appear in the Act and, therefore, a detailed annotation of this section is unnecessary. One or two things ought, however, to be noted.

Agriculture. This definition is the same definition as is contained in s.96 of the 1986 Act and reference should be made to Scammell & Denshams, *Law of Agricultural Holdings*, 7th Edition (Butterworths) and Muir Watt, *Agricultural Holdings*, 13th Edition (Sweet & Maxwell) for a detailed analysis.

Livestock. This definition is different from the definition in the 1986 Act which also includes any creature kept for the carrying on in relation to land of any agricultural activity. It is arguable that those words in the 1986 Act were superfluous, as they would be in this Act, as the farming of land includes references to the carrying on in relation to land of any agricultural activity in any event (see subs. (2)).

Index of defined expressions

39. In this Act the expressions listed below are defined by or otherwise fall to be construed in accordance with the provisions indicated—

agriculture, agricultural	section 38(1)
begins, beginning (in relation to a tenancy)	section 38(4)
building	section 38(1)
farm business tenancy	section 1
farming (of land)	section 38(2)
fixed term tenancy	section 38(1)
grant (of a tenancy)	section 38(3)
holding (in relation to a farm business tenancy)	section 38(1)
landlord	section 38(1) and (5)
livestock	section 38(1)
planning permission (in Part III)	section 27
provision (of a tenant's improvement) (in Part III)	section 15
the review date (in Part II)	section 10(2)
the RICS	section 38(1)
statutory review notice (in Part II)	section 10(1)
tenancy	section 38(1)
tenant	section 38(1) and (5)
tenant's improvement (in Part III)	section 15
termination (of a tenancy)	section 38(1).

Consequential amendments

40. The Schedule to this Act (which contains consequential amendments) shall have effect.

Short title, commencement and extent

41.—(1) This Act may be cited as the Agricultural Tenancies Act 1995.
(2) This Act shall come into force on 1st September 1995.
(3) Subject to subsection (4) below, this Act extends to England and Wales only.
(4) The amendment by a provision of the Schedule to this Act of an enactment which extends to Scotland or Northern Ireland also extends there, except that paragraph 9 of the Schedule does not extend to Northern Ireland.

Section 40

SCHEDULE

CONSEQUENTIAL AMENDMENTS

The Small Holdings and Allotments Act 1908 (c. 36)

1.—(1) Section 47 of the Small Holdings and Allotments Act 1908 (compensation for improvements) shall be amended as follows.
(2) In subsection (1), after "to any tenant" there shall be inserted "otherwise than under a farm business tenancy".
(3) In subsection (2), after "small holdings or allotments" there shall be inserted "otherwise than under a farm business tenancy".
(4) In subsection (3), after "if" there shall be inserted "he is not a tenant under a farm business tenancy and".
(5) In subsection (4), after "allotment" there shall be inserted "who is not a tenant under a farm business tenancy".
(6) After that subsection, there shall be inserted—
 "(5) In this section, 'farm business tenancy' has the same meaning as in the Agricultural Tenancies Act 1995."

The Law of Distress Amendment Act 1908 (c. 53)

2. In section 4(1) of the Law of Distress Amendment Act 1908 (exclusion of certain goods), for "to which that section applies" there shall be substituted "on land comprised in a tenancy to which that Act applies".

The Allotments Act 1922 (c. 51)

3. In section 3(7) of the Allotments Act 1922 (provision as to cottage holdings and certain allotments), after "landlord" there shall be inserted "otherwise than under a farm business tenancy (within the meaning of the Agricultural Tenancies Act 1995)".
4. In section 6(1) of that Act (assessment and recovery of compensation), after "contract of tenancy" there shall be inserted "(not being a farm business tenancy within the meaning of the Agricultural Tenancies Act 1995)".

The Landlord and Tenant Act 1927 (c. 36)

5. In section 17(1) of the Landlord and Tenant Act 1927 (holdings to which Part I applies), for the words from "not being" to the end there is substituted "not being—

(a) agricultural holdings within the meaning of the Agricultural Holdings Act 1986 held under leases in relation to which that Act applies, or
(b) holdings held under farm business tenancies within the meaning of the Agricultural Tenancies Act 1995."

6. In section 19(4) of that Act (provisions as to covenants not to assign etc. without licence or consent), after "the Agricultural Holdings Act 1986" there shall be inserted "which are leases in relation to which that Act applies, or to farm business tenancies within the meaning of the Agricultural Tenancies Act 1995".

The Agricultural Credits Act 1928 (c. 43)

7. In section 5(7) of the Agricultural Credits Act 1928 (agricultural charges on farming stock and assets) in the definition of "other agricultural assets", after "otherwise" there shall be inserted "a tenant's right to compensation under section 16 of the Agricultural Tenancies Act 1995,".

The Leasehold Property (Repairs) Act 1938 (c. 34)

8. In section 7(1) of the Leasehold Property (Repairs) Act 1938 (interpretation), at the end there shall be added "which is a lease in relation to which that Act applies and not being a farm business tenancy within the meaning of the Agricultural Tenancies Act 1995".

The Reserve and Auxiliary Forces (Protection of Civil Interests) Act 1951 (c. 65)

9.—(1) Section 27 of the Reserve and Auxiliary Forces (Protection of Civil Interests) Act 1951 (renewal of tenancy expiring during period of service or within two months thereafter) shall be amended as follows.

(2) In subsection (1), for the words from "are an agricultural holding" onwards there shall be substituted—
"(a) are an agricultural holding (within the meaning of the Agricultural Holdings Act 1986) held under a tenancy in relation to which that Act applies,
(b) are a holding (other than a holding excepted from this provision) held under a farm business tenancy, or
(c) consist of or comprise premises (other than premises excepted from this provision) licensed for the sale of intoxicating liquor for consumption on the premises."

(3) In subsection (5), after paragraph (b) there shall be inserted—
"(bb) the expressions 'farm business tenancy' and 'holding', in relation to such a tenancy, have the same meaning as in the Agricultural Tenancies Act 1995;".

(4) After that subsection, there shall be inserted—
"(5A) In paragraph (b) of the proviso to subsection (1) of this section the reference to a holding excepted from the provision is a reference to a holding held under a farm business tenancy in which there is comprised a dwelling-house occupied by the person responsible for the control (whether as tenant or servant or agent of the tenant) of the management of the holding."

(5) In subsection (6), for the words from the beginning to "liquor" there shall be substituted "In paragraph (c) of the proviso to subsection (1) of this section, the reference to premises excepted from the provision".

The Landlord and Tenant Act 1954 (c. 56)

10. In section 43(1) of the Landlord and Tenant Act 1954 (tenancies excluded from Part II)—
(a) in paragraph (a), for the words from "or a tenancy" to "1986" there shall be substituted "which is a tenancy in relation to which the Agricultural Holdings Act 1986 applies or a tenancy which would be a tenancy of an agricultural holding in relation to which that Act applied if subsection (3) of section 2 of that Act", and
(b) after that paragraph there shall be inserted—
"(aa) to a farm business tenancy;".

11. In section 51(1) of that Act (extension of Leasehold Property (Repairs) Act 1938), for paragraph (c) there shall be substituted—
"(c) that the tenancy is neither a tenancy of an agricultural holding in relation to which the Agricultural Holdings Act 1986 applies nor a farm business tenancy".

12. In section 69(1) of that Act (interpretation), after the definition of "development corporation" there shall be inserted—
> "'farm business tenancy' has the same meaning as in the Agricultural Tenancies Act 1995;".

The Opencast Coal Act 1958 (c. 69)

13.—(1) Section 14 of the Opencast Coal Act 1958 (provisions as to agricultural tenancies in England and Wales) shall be amended as follows.

(2) In subsection (1)(b), for "or part of an agricultural holding" there shall be substituted "held under a tenancy in relation to which the Agricultural Holdings Act 1986 (in this Act referred to as 'the Act of 1986') applies or part of such an agricultural holding".

(3) In subsection (2), for the words from "Agricultural" to "of 1986")" there shall be substituted "Act of 1986".

14. After section 14A of that Act, there shall be inserted—

"Provisions as to farm business tenancies

14B.—(1) Without prejudice to the provisions of Part III of this Act as to matters arising between landlords and tenants in consequence of compulsory rights orders, the provisions of this section shall have effect where—
   (a) opencast planning permission has been granted subject to a restoration condition, and
   (b) immediately before that permission is granted, any of the land comprised therein consists of the holding or part of the holding held under a farm business tenancy,

whether any of that land is comprised in a compulsory rights order or not.

(2) For the purposes of section 1 of the Agricultural Tenancies Act 1995 (in this Act referred to as 'the Act of 1995'), the land shall be taken, while it is occupied or used for the permitted activities, to be used for the purposes for which it was used immediately before it was occupied or used for the permitted activities.

(3) For the purposes of the Act of 1995, nothing done or omitted by the tenant or by the landlord under the tenancy by way of permitting any of the land in respect of which opencast planning permission has been granted to be occupied for the purpose of carrying on any of the permitted activities, or by way of facilitating the use of any of that land for that purpose, shall be taken to be a breach of any term or condition of the tenancy, either on the part of the tenant or on the part of the landlord.

(4) In determining under subsections (1) and (2) of section 13 of the Act of 1995 the rent which should be properly payable for the holding, in respect of any period for which the person with the benefit of the opencast planning permission is in occupation of the holding, or of any part thereof, for the purpose of carrying on any of the permitted activities, the arbitrator shall disregard any increase or diminution in the rental value of the holding in so far as that increase or diminution is attributable to the occupation of the holding, or of that part of the holding, by that person for the purpose of carrying on any of the permitted activities.

(5) In this section 'holding', in relation to a farm business tenancy, has the same meaning as in the Act of 1995.

(6) This section does not extend to Scotland."

15.—(1) Section 24 of that Act (tenant's right to compensation for improvements and other matters) shall be amended as follows.

(2) In subsection (1)(a), after "holding" there shall be inserted "held under a tenancy in relation to which the Act of 1986 applies".

(3) In subsection (10), after "Scotland" there shall be inserted "the words 'held under a tenancy in relation to which the Act of 1986 applies' in subsection (1)(a) of this section shall be omitted and".

16. After section 25 of that Act, there shall be inserted—

"Tenant's right to compensation for improvements etc.: farm business tenancies

25A.—(1) The provisions of this section shall have effect where—
   (a) any part of the land comprised in a compulsory rights order is held, immediately before the date of entry, under a farm business tenancy;
   (b) there have been provided in relation to the land which is both so comprised and so held ('the tenant's land') tenant's improvements in respect of which, immediately before that date, the tenant had a prospective right to compensation under section 16 of the Act of 1995 on quitting the holding on the termination of the tenancy;
   (c) at the end of the period of occupation, the tenant's land has lost the benefit of any such improvement; and

(d) immediately after the end of that period, the tenant's land is comprised in the same tenancy as immediately before the date of entry, or is comprised in a subsequent farm business tenancy at the end of which the tenant is not deprived, by virtue of section 23(3) of that Act, of his right to compensation under section 16 of that Act in respect of any tenant's improvement provided during the earlier tenancy in relation to the tenant's land.

(2) For the purposes of subsection (1) of this section, subsection (2) of section 22 of the Act of 1995 (which requires notice to be given of the intention to make a claim) shall be disregarded.

(3) Subject to subsection (4) of this section, Part III of the Act of 1995 shall apply as if—
  (a) the tenant's land were in the state in which it was immediately before the date of entry, and
  (b) the tenancy under which that land is held at the end of the period of occupation had terminated immediately after the end of that period and the tenant had then quitted the holding.

(4) Where the tenant's land has lost the benefit of some tenant's improvements but has not lost the benefit of all of them, Part III of the Act of 1995 shall apply as mentioned in subsection (3) above, but as if the improvements of which the tenant's land has not lost the benefit had not been tenant's improvements.

(5) For the purposes of subsections (1) and (4) of this section, the tenant's land shall be taken to have lost the benefit of a tenant's improvement if the benefit of that improvement has been lost (wholly or in part) without being replaced by another improvement of comparable benefit to the land.

(6) In this section 'holding', in relation to a farm business tenancy, 'tenant's improvement', 'termination', in relation to a tenancy, and references to the provision of a tenant's improvement have the same meaning as in the Act of 1995.

(7) This section does not extend to Scotland."

17.—(1) Section 26 of that Act (compensation for short-term improvements and related matters) shall be amended as follows.

(2) In subsection (1), after "agricultural land" there shall be inserted "and was not comprised in a farm business tenancy".

(3) In subsection (6), after "Scotland" there shall be inserted—
  "(za) in subsection (1) of this section, the words 'and was not comprised in a farm business tenancy' shall be omitted;".

18.—(1) Section 28 of that Act (special provision as to market gardens) shall be amended as follows.

(2) In subsection (1), after "market garden" there shall be inserted "and was not comprised in a farm business tenancy."

(3) In subsection (6), after "Scotland" there shall be inserted "in subsection (1) of this section, the words 'and was not comprised in a farm business tenancy' shall be omitted; and".

19. In section 51 of that Act (interpretation) in subsection (1)—
  (a) after the definition of "the Act of 1986" there shall be inserted—
    "'the Act of 1995' means the Agricultural Tenancies Act 1995;" and
  (b) after the definition of "emergency powers" there shall be inserted—
    "'farm business tenancy' has the same meaning as in the Act of 1995;".

20.—(1) Schedule 7 to that Act (adjustments between landlords and tenants and in respect of mortgages and mining leases and orders) shall be amended as follows.

(2) After paragraph 1, there shall be inserted—
  "1A.—(1) The provisions of this paragraph shall have effect where—
    (a) paragraphs (a) and (b) of subsection (1) of section 25A of this Act apply, and
    (b) the farm business tenancy at the end of which the tenant could have claimed compensation for tenant's improvements terminates on or after the date of entry, but before the end of the period of occupation, without being succeeded by another such subsequent tenancy.

  (2) In the circumstances specified in sub-paragraph (1) of this paragraph, the provisions of Part III of the Act of 1995—
    (a) shall apply, in relation to the tenancy mentioned in that sub-paragraph, as if, at the termination of that tenancy, the land in question were in the state in which it was immediately before the date of entry, and
    (b) if the tenant under that tenancy quitted the holding before the termination of his tenancy, shall so apply as if he had quitted the holding on the termination of his tenancy.

  (3) In sub-paragraph (2) of this paragraph, 'holding', in relation to a farm business tenancy, and 'termination', in relation to a tenancy, have the same meaning as in the Act of 1995."

(3) In paragraph 2, in sub-paragraph (1), after "agricultural holding" there shall be inserted "held under a tenancy in relation to which the Act of 1986 applies".

(4) After that paragraph there shall be inserted—

"2A.—(1) The provisions of this paragraph shall have effect where land comprised in a farm business tenancy is comprised in a compulsory rights order (whether any other land is comprised in the holding, or comprised in the order, or not), and—
  (a) before the date of entry there had been provided in relation to the land in question tenant's improvements (in this paragraph referred to as 'the former tenant's improvements') in respect of which, immediately before that date, the tenant had a prospective right to compensation under section 16 of the Act of 1995 on quitting the holding on the termination of the tenancy, and
  (b) at the end of the period of occupation the circumstances are such that Part III of that Act would have applied as mentioned in subsections (3) and (4) of section 25A of this Act, but for the fact that the benefit of the former tenant's improvements has been replaced, on the restoration of the land, by other improvements (in this paragraph referred to as 'the new improvements') of comparable benefit to the land.

(2) In the circumstances specified in sub-paragraph (1) of this paragraph, Part III of the Act of 1995 shall have effect in relation to the new improvements as if those improvements were tenant's improvements.

(3) Subsections (2) and (6) of section 25A of this Act shall apply for the purposes of this paragraph as they apply for the purposes of that section."

(5) After paragraph 3 there shall be inserted—

"3A. Where by virtue of section 25A of this Act a tenant is entitled to compensation for tenant's improvements as mentioned in that section and—
  (a) after the end of the period of occupation expenses are incurred in replacing the benefit of the tenant's improvements by other improvements of comparable benefit to the land, and
  (b) the person incurring those expenses (whether he is the landlord or not) is entitled to compensation in respect of those expenses under section 22 of this Act,
section 13 of the Act of 1995 shall apply as if the works in respect of which those expenses are incurred were not tenant's improvements, if apart from this paragraph they would constitute such improvements."

(6) At the end of paragraph 4, there shall be added—

"(7) In this paragraph 'agricultural holding' does not include an agricultural holding held under a farm business tenancy."

(7) After that paragraph there shall be inserted—

"4A.—(1) The provisions of this paragraph shall apply where—
  (a) immediately before the operative date of a compulsory rights order, any of the land comprised in the order is subject to a farm business tenancy, and
  (b) that tenancy continues until after the end of the period of occupation.

(2) The landlord or tenant under the tenancy may, by notice in writing served on his tenant or landlord, demand a reference to arbitration of the question whether any of the terms and conditions of the tenancy (including any term or condition relating to rent) should be varied in consequence of any change in the state of the land resulting from the occupation or use of the land in the exercise of rights conferred by the order; and subsection (3) of section 28 of the Act of 1995 shall apply in relation to a notice under this sub-paragraph as it applies in relation to a notice under subsection (2) of that section.

(3) On a reference by virtue of this paragraph, the arbitrator shall determine what variations (if any) should be made in the terms and conditions of the tenancy, and the date (not being earlier than the end of the period of occupation) from which any such variations are to take effect or to be treated as having taken effect; and as from that date the tenancy shall have effect, or, as the case may be, shall be treated as having had effect, subject to any variations determined by the arbitrator under this paragraph.

(4) The provisions of this paragraph shall not affect any right of the landlord or the tenant, or the jurisdiction of the arbitrator, under Part II of the Act of 1995; but where—
  (a) there is a reference by virtue of this paragraph and a reference under Part II of that Act in respect of the same tenancy, and
  (b) it appears to the arbitrator that the reference under Part II of that Act relates wholly or mainly to the consequences of the occupation or use of the land in the exercise of rights conferred by the order,
he may direct that proceedings on the two references shall be taken concurrently."

(8) In paragraph 5(1), after "agricultural holding" there shall be inserted "held under a tenancy in relation to which the Act of 1986 applies".

(9) In paragraph 6—
(a) in sub-paragraph (1), for "an agricultural holding" there shall be substituted "—
   (a) an agricultural holding held under a tenancy in relation to which the Act of 1986 applies, or
   (b) a holding under a farm business tenancy,"; and
(b) after sub-paragraph (2) there shall be added—
   "(2A) In sub-paragraph (1) of this paragraph, 'holding', in relation to a farm business tenancy, has the same meaning as in the Act of 1995."
(10) In paragraph 7—
(a) after "The provisions of" there shall be inserted "sub-paragraphs (1) to (6) of";
(b) for "that paragraph" there shall be substituted "those sub-paragraphs"; and
(c) after "subject to a mortgage" there shall be inserted "but not comprised in a farm business tenancy".
(11) After that paragraph there shall be inserted—
   "7A. The provisions of paragraph 4A of this Schedule shall apply in relation to mortgages of land comprised in farm business tenancies as they apply in relation to such tenancies, as if any reference in that paragraph to such a tenancy were a reference to such a mortgage, and any reference to a landlord or to a tenant were a reference to a mortgagee or to a mortgagor, as the case may be."
(12) In paragraph 12(1)(a), for the words from "did" to "holding" there shall be substituted "was not comprised in a tenancy in relation to which the Act of 1986 applies or in a farm business tenancy".
(13) In paragraph 13, after "or to a tenancy" there shall be inserted "(other than a reference to a tenancy in relation to which the Act of 1986 applies or a farm business tenancy)".
(14) In paragraph 25—
(a) in sub-paragraph (a), at the beginning there shall be inserted "subject to sub-paragraphs (ba), (bc), (bd)(i) and (be) of this paragraph,";
(b) after sub-paragraph (b), there shall be inserted—
   "(ba) in sub-paragraph (1) of paragraph 2, the words 'held under a tenancy in relation to which the Act of 1986 applies' shall be omitted;
   (bb) sub-paragraph (7) of paragraph 4 shall be omitted;
   (bc) in sub-paragraph (1) of paragraph 5, the words 'held under a tenancy in relation to which the Act of 1986 applies' shall be omitted;
   (bd) in paragraph (6)—
      (i) for paragraphs (a) and (b) of sub-paragraph (1) there shall be substituted the words 'an agricultural holding'; and
      (ii) sub-paragraph (2A) shall be omitted;
   (be) in sub-paragraph (1)(a) of paragraph 12, for the words 'was not comprised in a tenancy in relation to which the Act of 1986 applies or in a farm business tenancy' there shall be substituted the words 'did not constitute or form part of an agricultural holding';" and
(c) in sub-paragraph (c), for "7" there shall be substituted "1A, 2A, 3A, 4A, 7, 7A".

The Agriculture (Miscellaneous Provisions) Act 1963 (c. 11)

21.—(1) Section 22 of the Agriculture (Miscellaneous Provisions) Act 1963 (allowances to persons displaced from agricultural land) shall be amended as follows.
(2) In subsection (1), for paragraph (a) there shall be substituted—
   "(a) the land—
      (i) is used for the purposes of agriculture (within the meaning of the Agricultural Tenancies Act 1995) and is so used by way of a trade or business, or
      (ii) is not so used but is comprised in a farm business tenancy (within the meaning of the Agricultural Tenancies Act 1995) and used for the purposes of a trade or business,".
(3) In subsection (6)(c), for "the Agricultural Holdings Act 1986" there shall be substituted ", the Agricultural Tenancies Act 1995".

The Leasehold Reform Act 1967 (c. 88)

22. In section 1(3) of the Leasehold Reform Act 1967 (tenants entitled to enfranchisement or extension), for paragraph (b) there shall be substituted—

"(b) it is comprised in—
   (i) an agricultural holding within the meaning of the Agricultural Holdings Act 1986 held under a tenancy in relation to which that Act applies, or
   (ii) the holding held under a farm business tenancy within the meaning of the Agricultural Tenancies Act 1995."

The Agriculture (Miscellaneous Provisions) Act 1968 (c. 34)

23. In section 12 of the Agriculture (Miscellaneous Provisions) Act 1968 (additional payments in consequence of compulsory acquisition etc of agricultural holdings), after subsection (1) there shall be inserted—
"(1A) No sum shall be payable by virtue of subsection (1) of this section in respect of any land comprised in a farm business tenancy within the meaning of the Agricultural Tenancies Act 1995."

The Land Compensation Act 1973 (c. 26)

24. In section 48 of the Land Compensation Act 1973 (compensation in respect of agricultural holdings) at the beginning of subsection (1) there shall be inserted "Subject to subsection (1A) below" and after subsection (1) there shall be inserted—
"(1A) This section does not have effect where the tenancy of the agricultural holding is a tenancy to which, by virtue of section 4 of the Agricultural Tenancies Act 1995, the Agricultural Holdings Act 1986 does not apply."

The Rent (Agriculture) Act 1976 (c. 80)

25.—(1) Section 9 of the Rent (Agriculture) Act 1976 (effect of determination of superior tenancy, etc) shall be amended as follows.
(2) In subsection (3), after "the Agricultural Holdings Act 1986" there shall be inserted "held under a tenancy in relation to which that Act applies and land comprised in a farm business tenancy within the meaning of the Agricultural Tenancies Act 1995."
(3) In subsection (4), for the words from "or" at the end of paragraph (b) onwards there shall be substituted—
"(c) a tenancy of an agricultural holding within the meaning of the Agricultural Holdings Act 1986 which is a tenancy in relation to which that Act applies; or
(d) a farm business tenancy within the meaning of the Agricultural Tenancies Act 1995."
26. In Schedule 2 to that Act (meaning of "relevant licence" and "relevant tenancy"), in paragraph 2 for the words from "and a tenancy" to the end there shall be substituted ", a tenancy of an agricultural holding within the meaning of the Agricultural Holdings Act 1986 which is a tenancy in relation to which that Act applies, and a farm business tenancy within the meaning of the Agricultural Tenancies Act 1995."

The Rent Act 1977 (c. 42)

27. For section 10 of the Rent Act 1977 there shall be substituted—

"**Agricultural holdings etc.**
10.—(1) A tenancy is not a protected tenancy if—
(a) the dwelling-house is comprised in an agricultural holding and is occupied by the person responsible for the control (whether as tenant or as servant or agent of the tenant) of the farming of the holding, or
(b) the dwelling-house is comprised in the holding held under a farm business tenancy and is occupied by the person responsible for the control (whether as tenant or as servant or agent of the tenant) of the management of the holding.
(2) In subsection (1) above—
  'agricultural holding' means any agricultural holding within the meaning of the Agricultural Holdings Act 1986 held under a tenancy in relation to which that Act applies, and
  'farm business tenancy', and 'holding' in relation to such a tenancy, have the same meaning as in the Agricultural Tenancies Act 1995."
28.—(1) Section 137 of that Act (effect on sub-tenancy of determination of superior tenancy) shall be amended as follows.
(2) In subsection (3), after "the Agricultural Holdings Act 1986" there shall be inserted "held under a tenancy to which that Act applies and land comprised in a farm business tenancy within the meaning of the Agricultural Tenancies Act 1995."

(3) In subsection (4), in paragraph (c), for the words from "applies" onwards there shall be substituted "applies—
(i) a tenancy of an agricultural holding within the meaning of the Agricultural Holdings Act 1986 which is a tenancy in relation to which that Act applies, or
(ii) a farm business tenancy within the meaning of the Agricultural Tenancies Act 1995."

The Protection from Eviction Act 1977 (c. 43)

29. In section 8(1) of the Protection from Eviction Act 1977 (interpretation)—
(a) in paragraph (d), after "Agricultural Holdings Act 1986" there shall be inserted "which is a tenancy in relation to which that Act applies", and
(b) at the end there shall be added—
"(g) a farm business tenancy within the meaning of the Agricultural Tenancies Act 1995."

The Housing Act 1985 (c. 68)

30. In Schedule 1 to the Housing Act 1985 (tenancies which are not secure tenancies), for paragraph 8 there shall be substituted—

"*Agricultural holdings etc.*

8.—(1) A tenancy is not a secure tenancy if—
(a) the dwelling-house is comprised in an agricultural holding and is occupied by the person responsible for the control (whether as tenant or as servant or agent of the tenant) of the farming of the holding, or
(b) the dwelling-house is comprised in the holding held under a farm business tenancy and is occupied by the person responsible for the control (whether as tenant or as servant or agent of the tenant) of the management of the holding.
(2) In sub-paragraph (1) above—
'agricultural holding' means any agricultural holding within the meaning of the Agricultural Holdings Act 1986 held under a tenancy in relation to which that Act applies, and
'farm business tenancy', and 'holding' in relation to such a tenancy, have the same meaning as in the Agricultural Tenancies Act 1995."

The Landlord and Tenant Act 1985 (c. 70)

31. In section 14(3) of the Landlord and Tenant Act 1985 (leases to which section 11 does not apply), at the end there shall be added "and in relation to which that Act applies or to a farm business tenancy within the meaning of the Agricultural Tenancies Act 1995."

The Agricultural Holdings Act 1986 (c. 5)

32. In Schedule 6 to the Agricultural Holdings Act 1986 (eligibility to apply for a new tenancy under Part IV of that Act), in paragraph 6 (occupation to be disregarded for purposes of occupancy condition), in sub-paragraph (1) after paragraph (d) there shall be inserted—
"(dd) under a farm business tenancy, within the meaning of the Agricultural Tenancies Act 1995, for less than five years (including a farm business tenancy which is a periodic tenancy),".

The Housing Act 1988 (c. 50)

33. In section 101(2) of the Housing Act 1988 (which relates to tenancies and licences affecting property proposed to be acquired under Part IV of that Act), after "smallholdings)" there shall be inserted "nor the Agricultural Tenancies Act 1995 (farm business tenancies)".
34. In Schedule 1 to that Act (tenancies which cannot be assured tenancies), for paragraph 7 there shall be substituted—

"*Tenancies of agricultural holdings etc*

7.—(1) A tenancy under which the dwelling-house—
(a) is comprised in an agricultural holding, and
(b) is occupied by the person responsible for the control (whether as tenant or as servant or agent of the tenant) of the farming of the holding.
(2) A tenancy under which the dwelling-house—
(a) is comprised in the holding held under a farm business tenancy, and

(b) is occupied by the person responsible for the control (whether as tenant or as servant or agent of the tenant) of the management of the holding.

(3) In this paragraph—

'agricultural holding' means any agricultural holding within the meaning of the Agricultural Holdings Act 1986 held under a tenancy in relation to which that Act applies, and

'farm business tenancy' and 'holding', in relation to such a tenancy, have the same meaning as in the Agricultural Tenancies Act 1995."

The Town and Country Planning Act 1990 (c. 8)

35.—(1) Section 65 of the Town and Country Planning Act 1990 (notice etc. of applications for planning permissions) shall be amended as follows.

(2) In subsection (2), for "a tenant of any agricultural holding any part of which is comprised in that land" there shall be substituted "an agricultural tenant of that land".

(3) In subsection (8), for the definition of "agricultural holding" there shall be substituted—

"'agricultural tenant', in relation to any land, means any person who—

(a) is the tenant, under a tenancy in relation to which the Agricultural Holdings Act 1986 applies, of an agricultural holding within the meaning of that Act any part of which is comprised in that land; or

(b) is the tenant, under a farm business tenancy (within the meaning of the Agricultural Tenancies Act 1995), of land any part of which is comprised in that land;".

The Coal Mining Subsidence Act 1991 (c. 45)

36. In section 21 of the Coal Mining Subsidence Act 1991 (property belonging to protected tenants) in subsection (3), after paragraph (a) there shall be inserted—

"(aa) a tenant under a farm business tenancy within the meaning of the Agricultural Tenancies Act 1995;".

37. In Schedule 3 to that Act (property belonging to protected tenants) in paragraph 1(2), after paragraph (b) there shall be inserted—

"(bb) section 20 of the Agricultural Tenancies Act 1995;".

INDEX

**References are to the text of the Act and the General Note at the specified section or Schedule.
N indicates that the reference is to the General Note only**

AGREED SUCCESSIONS, s.4
AGRICULTURAL FIXTURES, s.8
AGRICULTURAL LANDS TRIBUNAL,
    succession tenancies, directions as to, s.4
AGRICULTURE,
    definition of, s.1
AGRICULTURE CONDITION,
    grazing, s.1
    lifestock, s.1
    meaning, s.1
    non-compliance with, s.1
    pesticide testing, s.1N
AGRICULTURE (MISCELLANEOUS PROVISIONS) ACT 1963,
    amendment, Sch.
AGRICULTURE (MISCELLANEOUS PROVISIONS) ACT 1968,
    amendment, Sch.
AGRICULTURAL CREDITS ACT 1928,
    amendment, Sch.
AGRICULTURAL FIXTURES, s.8N
AGRICULTURAL HOLDINGS ACT 1986,
    amendment, Sch.
    succession rules, s.4N
ALLOTMENTS ACT 1922,
    amendment, Sch.
ALTERNATIVE DISPUTE RESOLUTION, ss.28, 29
ANNUAL PERIODIC TENANCIES,
    notice to quit, ss.5,6
ARBITRATION,
    arbitrator. See ARBITRATOR
    award time limits, s.28N
    dispute resolution through, s.28N
    general provisions, s.30
    improvements, as to,
        consent refusal, where, s.19
        notice served, s.22
        planning permission, s.18
        right to, s.8
        settlement of claim, s.22
    rent review,
        appointment for, s.12
        factors taken into account, s.13
        fall back provisions, s.28N
    role of, s.13
    retrospective, s.19
    statutory, s.28
ARBITRATOR,
    appointment, s.19
    approval of, s.19
    considerations for, s.19

ARBITRATOR—*cont.*
    joint appointment, s.29
    powers of, s.19
    RICS appointment of, ss.22, 28

BEST RENT ESTIMATION, s.34N
BREAK CLAUSES, ss.7, 24
BUSINESS CONDITIONS,
    compliance, proof of, s.1
    meaning, s.1

CAPITAL MONEY,
    apply, power to, s.33
    income, repayment out of, s.33
    raise, power to, s.33
    repayment of, s.33
    settled land, in, s.33N
CHATTEL ASSETS, s.8N
COAL MINING SUBSIDENCE ACT 1991,
    amendment, Sch.
COMMERCIAL FARMING, s.1
COMMERCIAL TENANCIES, s.1
COMMODITIES,
    rent review linked to, s.9N
COMPENSATION,
    Evesham Custom, s.4
    improvements of tenant. *See* IMPROVEMENTS
CONSENT,
    improvements, to, ss.17, 19, 22N
CRIMINAL OFFENCES,
    deed preparation, s.35N
CROWN LAND, s.37

DEATH,
    relative of tenant, of, s.4
    tenant, of, s.4
DEED,
    agricultural tenancies executed by, s.35N
    criminal offences relating to, s.35N
DESIGNATIONS,
    intangible advantage, as, s.15N
DISPUTE RESOLUTION,
    arbitration, through, s.28
DISTURBANCE PAYMENTS, s.26N
DOCUMENTS,
    surveyors preparing, s.35
    valuers preparing, s.35

EVESHAM CUSTOM, s.4
EVIDENCE,
    rent review agreement, of, s.9N

Index

FARM BUSINESS TENANCY,
  agriculture conditions, s.1
  breach, use of land in, s.1
  business conditions, s.1
  compliance, proof of, s.1
  form of, s.1
  grazing agreements, s.1
  licences, informal, s.1
  meaning of, s.1
  non-agricultural use, s.1
  notice conditions of s.1. *See also* NOTICE
  rent review under. *See* RENT REVIEW
  short term lets replaced by, s.1
  surrender and re-grant of, s.3
  tenancies that cannot be, s.2
  term of, s.1N
  termination. *See* TERMINATION
  three years or more, for, s.1
  unlawful uses disregarded, s.1
  variation of terms, s.17
FARMED,
  requirement of, s.1
FIXTURES,
  agricultural, s.8
  annexation, degree of, s.8
  chattel assets, as, s.8
  damage to, s.8
  erection, s.8
  landlord purchasing, s.8
  notice of removal, s.8
  physical improvements as, s.15
  tenant's right to remove, s.8N
  title to, s.8N
  trade, s.8N

GOODWILL,
  intangible advantage, as, s.15N
GRASS KEEP,
  licences, s.1N
GRAZING,
  agriculture, as, s.1
  lets, s.1N

HOBBY FARMING, s.1N
HOLDING, s.23
HOUSING ACT 1985,
  amendment, Sch.

IMPROVEMENTS,
  compensation for,
    amount of, s.20
    capitalised rent values, s.20N
    double, s.23
    eligibility conditions, s.15
    enforcement, s.22
    landlord's consent required, s.17
    milk quotas, ss.15, 16, 17
    planning permission, ss.18, 20, 21

IMPROVEMENTS—*cont.*
  compensation for,—*cont.*
    possession resumed, s.24
    recovery of, s.26
    reduction in, s.15
    removal of improvements, s.16
    right to, s.16
    rolling-over, s.23N
    settlement of claims, s.22
    severance of reversionary estate, where, s.25
    succession tenancies, s.23
    tenant, to, s.8
    time limits, s.22
    unimplemented planning permission, s.20
    valuation basis, s.20
  consent to,
    conditions of, s.19
    generally, s.17
    time limit, s.19
  intangible advantages, s.15
  meaning, s.15
  physical, s.15
  removal by tenant, s.8
  rent review upon, s.13
  routine, ss.15, 17
INDEXATION,
  rent review, s.9N
INTANGIBLE ADVANTAGES, s.15

LAND AGENTS,
  tenancy agreement prepared by, s.35N
LAND COMPENSATION ACT 1973,
  amendment, Sch.
LANDLORD,
  improvements, consent given to, s.32
LANDLORD AND TENANT ACT 1927,
  amendment, Sch.
LANDLORD AND TENANT ACT 1954,
  amendment, Sch.
LANDLORD AND TENANT ACT 1985,
  amendment, Sch.
LAW OF DISTRESS AMENDMENT ACT 1908,
  amendment, Sch.
LEASEHOLD PROPERTY (REPAIRS) ACT 1938,
  amendment, Sch.
LEASEHOLD REFORM ACT 1967,
  amendment, Sch.
LICENCES,
  farm business tenancy conversion from, s.1N
  grass keep, s.1N
  intangible advantage, as, s.15
LIVESTOCK,
  agriculture, as, s.1
  meaning, s.38

MARKET GARDENS,
  Evesham Custom, s.4N

Index

MARKET GARDENS—*cont.*
  improvements to, s.4
  suitable holdings, s.4
MARRIAGE,
  tenant, of, s.7N
MILK QUOTAS,
  intangible advantage, as ss.15, 16, 17N
MINISTRY,
  consent licences, s.1N
  consent tenancies, s.1N
MORTGAGES OF AGRICULTURAL LAND,
  Law of Property Act 1925,
    amendment, s.31
  leasing powers, s.31
  pre-existing, s.31

NEW TENANCIES,
  application of law to, s.4
  surrender and re-grant through, s.4
NON-AGRICULTURAL USE,
  diversification into, s.1
NOTICE,
  conditions,
    compliance with, s.3
    generally, s.1
    surrender and re-grant, s.3
  counter-notice, s.6
  exchange of, s.1
  failure to serve, s.1
  form of, ss.1N, 5
  quit, to, s.6
  "relevant day" for, s.1
  service, s.36
  termination by, ss.5, 6
    part, possession resumed of, s.7

OPENCAST COAL ACT 1958,
  amendment, Sch.
OPEN MARKET FORMULA, ss.12, 13
ORAL LETTINGS,
  authorisation of agreements for, s.31
  rent review, s.9

PERIODIC TENANCIES,
  annual, ss.5, 6
  monthly, s.6N
  quarterly, ss.6N, 7N
  weekly, s.6N
PESTICIDE TESTING, s.1N
PHYSICAL IMPROVEMENTS, s.15
PLANNING PERMISSION,
  compensation, ss.18, 21
  improvements, s.15
  intangible advantages, s.18
  landlord's consent to, s.18
  possession resumed, where, s.24
  unimplemented, s.20

PONY PADDOCKS, s.1
POSSESSION,
  resumed, s.24
PROTECTION FROM EVICTION ACT 1977,
  amendment, Sch.

QUIT, NOTICE TO, s.6N

RE-ENTRY RIGHT, s.5
RENT,
  best, estimation of, s.34N
  review. *See* RENT REVIEW
RENT ACT 1977,
  amendment, Sch.
RENT (AGRICULTURE) ACT 1976,
  amendment, Sch.
RENT REVIEW,
  agreed basis, s.12
  arbitration, ss.10, 12, 13, 28
  clauses, s.9
  commodity prices, linked to, s.9N
  date of, s.10
  farm, moving, s.12N
  formula, s.9
  frequency of, ss.9, 10
  indexation, s.9
  minor alterations, s.10
  notice requiring, s.10
  provision for, s.4
  reversion severance, following, s.11
  stages in, s.9
  statutory, s.9
  three-yearly, ss.10, 11
  timing of, s.9
  triggering mechanisms, ss.9, 10
RESERVE AND AUXILIARIES FORCES (PROTECTION OF CIVIL INTEREST) ACT 1951,
  amendment, Sch.
RETIREMENT,
  tenant, of, s.4
ROUTINE IMPROVEMENTS, ss.15, 19, 26N

SERVICE,
  notice, methods for, s.36
SEVERANCE,
  reversionary estate, ss.5N, 11, 25
SHORT TERM TENANCIES,
  rent review, s.9
SMALL HOLDINGS AND ALLOTMENTS ACT 1908,
  amendment, Sch.
SUBSIDIES, s.15
SUCCESSION TENANCIES,
  improvements, compensation for, s.23
  meaning, s.4N

Index

SURRENDER AND RE-GRANT, ss.3, 4
SURVEYORS,
    documents prepared by, s.35N

TENANCY,
    defined, s.1
TENANT,
    buildings removed by, s.8
    death of, s.4N
    fixtures removed by, s.8
    improvements by. *See* IMPROVEMENTS
    marriage of, s.7
    moving farms, s.4N
    retirement of, s.4N
    severance deed, party to, s.11
TERMINATION,
    farm business tenancy,
        notice to quit, s.5
        two years or more, after, s.5
    improvements incomplete on, s.18
TOWN AND COUNTRY PLANNING ACT 1990,
    amendment, Sch.
TRADE FIXTURES, s.8N

VALUERS,
    documents prepared by, s.35N

WORDS AND PHRASES,
    accredited person, s.35
    agreed succession, s.4
    agricultural, s.38
    agricultural holding, s.2
    agriculture, s.38
    beginning, s.38
    building, s.38
    capital money, s.33
    Crown interest, s.37
    farm business tenancy, s.1
    farmed, s.1
    farming, s.38
    fixed equipment, s.19
    fixed term tenancy, s.38
    holding, s.38
    landlord, s.38
    livestock, s.38
    original holding, s.24
    planning permision, s.27
    primarily or wholly agricultural, s.1
    relevant development, s.21
    review date, s.10
    RICS, s.38
    routine improvement, s.19
    settled land, s.33
    statutory review notice, s.10
    tenancy, s.38
    tenant, s.38
    tenant's improvement, s.15
    termination date, s.24